WAR AT HOME

WAR AT HOME

A Call to Reform the Military Justice System
and to Give Voice to the Ones Who Need to

BREAK THE SILENCE,

DESTROY THE STIGMA and

END THE CYCLE OF ABUSE

JAKIA LINDLEY

MOVING MOUNTAINS
PUBLISHING

<u>www.jmlindley.com</u>

First Edition

ISBN: Paperback 979-8-9860213-1-7
ISBN: Hardcover 979-8-9860213-2-4
ISBN: Electronic 979-8-9860213-0-0

Library of Congress Control Number: 2022906945

Edited by Scott Carbonara and Jocelyn Carbonara
Cover design by Rafael Andres
Book designed by Rodney Miles, www.rodneymiles.com
Publishing Consultant PRESStinely

Contents

"You may not control all the events that happen to you, but you can decide not to be reduced by them."

~ Maya Angelou

For my daughter.

No matter the cards you are dealt, no matter what obstacles you may face, know that you can overcome every adversity that comes your way. Never dim your light so that others may shine. Never quiet your voice so that others may be heard. The world is big, but never let the world make you feel small. You are capable. You are strong. You are beautiful. You are loved.

Mommy loves you forever and in all my days.

Acknowledgements

IT'S AN UNDERSTATEMENT to say I wouldn't be here without the incredible support from my family and friends. Mom, I could write an entire book on your unyielding love and support. My siblings, you all have been my best friends from day one. I couldn't have asked for a better set of partners in crime to go through this thing called life. Dad and Stepmom, thank you for always being there for me, day or night.

Because of you all, I am still here. Because of you, I no longer face war at home.

A Note to Readers

THIS BOOK WAS written in 2020, and in just a few short years, some of the changes I recommended have been already implemented. Through advocacy work all across the world, our combined voices and activism have already resulted in major changes, such as the passage of the I Am Vanessa Guillén Act to help and protect victims of sexual violence in the military[1], with many more on the way.

We have come so far in just this short time, but there is still so much work to be done.

---

[1] https://www.nbcnews.com/news/latino/vanessa-guillen-act-brings-historic-military-reform-calls-justice-pers-rcna12952, and for more info on the law, please see: https://www.congress.gov/bill/116th-congress/house-bill/8270?s=1&r=19

Foreword

WHEN I READ Jakia's story, my heart was gripped. I saw her beauty—both in her strength as a woman and courage to become the leader of her own destiny. I felt her pain—in her abuse, her objectification, and the chipping away at her reality by those who didn't truly see her. And I wanted to come alongside her and add to her power—as we need a voice like Jakia's to remind us that we are not alone, and we can thrive.

Jakia endured sexual abuse and assault in her childhood. A stranger identified her as *something special*. As she grew older, both men and women objectified her, heaping unwanted sexual attention upon her—when all she wanted was to do her job. She faced discrimination, harassment, and abuse from those in power. She suffered abuse from some of the people closest to her. And somehow, she survived, facing her battles head on, and giving voice to those who could not speak for themselves.

I also couldn't help but relate some parts of her story to my own. Before I became an actress and model, I was discovered by a photographer at just thirteen years old while walking on the Santa Monica Pier with my mother and brother. I was flattered, and I felt important—and beautiful.

But the truth is that photographer didn't discover me; he discovered my appearance.

"She's got the right look," he told my mom as he convinced her to let me take part in a photoshoot for a famous designer.

No, that photographer never got to know me, the person beneath my skin. From that first photoshoot, few took an interest in discovering the real me. Instead, they sought me out to serve as a living, breathing, walking clothing hanger for designers across the globe. At just thirteen, I became an object, nothing more.

No matter where I went—walking the runways of Milan, gracing the cover of *Elle* magazine at age fourteen (the youngest ever to do so), shooting music videos for Tupac Shakur and Mick Jagger, or starring in movies like *Blow* and *Vanilla Sky*—no one really knew me. Like Jakia, I wore a mask and played a role. Both of us had people try to get close to us as a way to take advantage of us. And both of us, for a time, lost our voices, our power, our mental health, and nearly our lives.

So yes, I relate to Jakia and what she endured at the hands of others. But unfortunately, we are not alone. Countless unnamed victims suffer from similar stories of objectification, sexual abuse, assault, domestic violence, harassment, discrimination, and depression from their experiences.

One in five women in the US will be raped in their lifetime. Since COVID-19, domestic violence has skyrocketed, resulting in nearly 700,000 reported cases and more than 35,000 deaths in the US alone. Victims of domestic abuse have an increased risk for depression, suicidal behaviors, and addiction. More than 80 percent of women and 40 percent of men have experienced sexual harassment. Of those harassed at work, more than 70 percent don't report it, fearing retribution.

As a woman serving in the military, Jakia faced additional risks and challenges in a male-dominated culture. Her inward and outward beauty attracted stalkers and predators alike. As a woman of color, she met with discrimination and outright racism.

Her story is riveting, powerful, and necessary. She takes us on her journey from childhood through adulthood, experiencing childhood assault, domestic abuse, sexual assault, and utter despair. While her spirit wavered at times, her light never went out. Today, she stands proudly with strength, showing others that fighting back and healing starts with something as seemingly small as standing up for ourselves.

While Jakia's story is unique, readers will see a bit of themselves on each page. Because hers is the story of every innocent child, every daughter, every sister, every member of the military, everyone who has dreamed, every person of color, and every human threatened by crushing circumstances. And my prayer is that you will use the example of her strength to secure safety if you are not safe today. And if you are safe, I pray you will stand up against abuse of any kind until the world is safe for every member.

Jakia makes herself vulnerable in the raw, honest retelling of the wars she's been through, the wars no one wants to talk about: domestic violence, sexual abuse, discrimination, and abuse of power. Throughout, she displays a maturity and courage beyond her years. She bravely tells how she finally broke the silence, destroyed the stigma, and ended the cycle of abuse—while rediscovering herself.

~**Jennifer Gimenez**
American supermodel, actress, reality television personality,
and addiction recovery advocate

Introduction

I did not recognize ABUSE when it first appeared,

because I could not imagine one soul hurting another.

I did not retreat from ABUSE when it grew;

rather, I washed my face, readied my smile, and wore camouflage armor.

I did not run from ABUSE when it persisted,

choosing instead to blame myself and to see the best in you.

When I finally stood strong against ABUSE,

it was not because I hated you.

I did it because I realized, at last, I deserved better.

AS AN AIRMAN in the United States Air Force, I endured years of ongoing sexual harassment and assault from both male and female colleagues, and I did my best to shut my mouth, suck it up, and be a strong airman.

But on June 30, 2020, investigators recovered the lifeless body of US Army Specialist (SPC) Vanessa Guillen, and I knew it was time to break my silence.

I never met SPC Vanessa Guillen, but I felt like we shared a connection that transcended geography, ethnicity, and branch of service. My heart ached for her story, wondering what kind of special hell she endured at the hands of a fellow soldier.

All military recruits recite the "Oath of Enlistment" at a solemn event where they promise to "support and defend the Constitution of the United States against all enemies, foreign and domestic." The murder of SPC Guillen occurred while she did her job domestically. No foreign enemy touched her.

1

I have faithfully served my country in the Air Force for more than eleven years. And like SPC Guillen, I never met foes as savage as fellow airmen, chiefs, and commanders stateside in the United States Air Force.

"That could have been me," I told my husband and closest friends when I learned of SPC Guillen's murder.

SPC Guillen served in the United States Army at Fort Hood in Texas, but she was not the only domestic casualty at Fort Hood in 2020. From January through the end of August in 2020, twelve other soldiers were murdered, went missing, or died in accidents.

If that number doesn't shock you, know that Fort Hood is just one military base. Between the Army, Navy, Marines, Air Force and Space Force, the US military operates more than 350 domestic installations in the fifty states and Puerto Rico.

How many active-duty service personnel, I wondered, *show up each day to serve this country, but instead of appreciation they endure hazing, sexual harassment, and abuse? How many die at the hands of fellow service personnel like SPC Vanessa Guillen did?*

How many take their own lives, unable to find relief from ongoing, uninvited abuse from within their branch of service? Suicide. Suicide is the second leading cause of death for active military members. In almost every military death by suicide, warning signs (like increased alcohol consumption, depression, and other behavioral changes) went unaddressed.

How many service members struggle with mental health issues? Nearly 60 percent of military veterans treated by the Veterans Health Administration suffered from mental health conditions (Va.gov, 2020). To lower the suicide rate, we must address the underlying causes.

My name is Jakia Monee Lindley. For years I kept my secrets quiet. But the time has come for me to speak. No, I cannot speak for every member of the service, nor will I try to. But I will tell you about the unresolved trauma I brought into my own military service. I will detail the harassment I faced as a woman in the Air Force, and the abuse I suffered when I obeyed the order of "if you see something, say something." I will share about the domestic violence I experienced at the hands of my ex-husband, a fellow airman. I will be open about the system that sucked the will to live from me.

Trauma from abuse can crush your spirit unless you confront your enemies. After years of stuffing my emotions, putting on a happy face, and enduring hardship like a good airman, I realized I needed to stand toe-to-toe against my foes. I needed to fight back.

This book is dedicated to the countless marines, soldiers, sailors, airmen, and guardians that suffer in silence. This is for all of you.

It's time that we demand better.

Note: In sharing my story, *War at Home*, I relived several painful personal experiences in hopes of breaking the silence, destroying the stigma, and ending the cycle of abuse that is too common for those in toxic, abusive relationships. I believe that by working together, we can forge a stronger, more positive culture, one that makes the world safer for us as well as future generations. Since it has never been my intention to embarrass or single out any individual, I changed the names and many locations in my story to preserve their anonymity.

"I swore never to be silent whenever and wherever human beings endure suffering and humiliation. We must take sides. Neutrality helps the oppressor, never the victim. Silence encourages the tormentor, never the tormented."

~ Elie Wiesel

Chapter 1

The Old Soul

"Old souls are usually childlike in many ways, having the playfulness and simplicity of children, while maintaining a certain world-weariness and deep insight."

~ Mateo Sol

I PUMPED MY little legs up and down on the pedals of my bike until I reached the top of the steep hill. Once my older sisters and cousins caught up with me, we all raced to the bottom. As the youngest in the group, only recently out of training-wheels, I took great pride in being so fast. But our competition wasn't just about speed. To be declared the "winner," you had to be the last person to put on the brakes, which required fearlessness. After winning yet another race, my grandma took the shine off my achievement as she screamed at me from her front porch.

"Kia, stop playing in the street! You trying to get hit by a car?" she shouted for the entire neighborhood to hear.

As my face flushed red, I turned back to see my sisters and cousins keeled over in the grass, laughing at me. That was me, always pushing my limits, and apparently, my grandma's limits as well.

Thinking back on those days, I long for the simplicity of what it took to be deemed *the winner*—as well as the gentle consequences of pushing things too far, as defined by my grandma. As years passed, I never lost that desire to push myself. I did learn, though, that some of life's consequences hurt more than others. And I discovered that some people play by their own rules, choosing to dish out pain to those who cross their paths.

I was born in Memphis, Tennessee, in the spring of 1992. Since my mom and grandma moved us to Ohio shortly after I was born, I have very few memories there as a child. As an adult, I find myself with a nostalgic yearning for Tennessee, and the state of my birth has become my favorite vacation spot.

Where I grew up, you either labored in a factory doing blue-collar work, or you tried your luck at becoming a professional athlete. I wish I had a nickel for every time someone asked me, "Do you know LeBron James?" I'd have even more nickels if I counted the occasions when someone said, "So you must be a huge Cavaliers fan!" Once people learn that I'm not from Cleveland, Columbus, or Cincinnati, their faces take on a look that says, "But didn't you just say you were from Ohio?"

Yep, Ohio has other cities besides the three big Cs. My adopted home state is home of the Buckeyes and their infamous rivalry with Michigan State. And yes, I went to Amish Country for my sixth-grade field trip. Ohio is the state where one is most likely to get caught up in rain while wearing flip-flops, shorts, and a light jacket—only to be followed by blistering heat before watching nine inches of snow fall—all before you can finish your morning cup of coffee.

For all the random things that Ohio has to offer, I would be remiss if I didn't mention the amazing things that make the state unique. At any given time, no matter where you are, the moment you yell out "O-H" someone will immediately shout back "I-O." You know you're from Ohio if you have at least one photo of you and three friends using your bodies to spell out O-H-I-O. Both traditions started when several rabid Ohio State fans served in the Pacific theater in the US Navy during 1942. They created their own sea shanty, borrowing the tune of "Row, Row, Row Your Boat" but spelling out O-H-I-O-S-T-A-T-E. After the end of the Second World War, one of those sailors attended Ohio State and joined the cheerleading squad, where he taught the cheer and gesture to the fans.

I take great pride in being from Ohio, but on any given day you can find some native complaining about our state while outlining their escape plan. Most plans include fleeing to Florida or one of the Carolinas to solve all their problems. Five years later, you can find that same person sitting in the same bar proclaiming, "I can't wait to move. I'm sick of this place," in between shots. Ohio is the place where only natives can trash-talk our home.

If you're wondering, I grew up in the relatively small town of Massillon, Ohio. Growing up in Massillon had its advantages, like living down the street from my Aunt Grace and grandma with tons of cousins just blocks away. Our families enjoyed cookouts every Sunday. On most holidays, we packed into Aunt Grace's house for get-togethers. And if I ever found myself in trouble, I had more than enough cousins to have my back. With the number of us, I'm sure we made up half the population.

But family can be a source of pain, too. At age four, after playing with every toy and imaginary friend I could concoct, boredom overtook me. Mom worked long hours, and when she wasn't working, she had a full-time job raising three girls on her own. One day after I'd exhausted all the fun I could create by myself, I begged my mom to let me play down at Aunt Grace's, a few houses away. Once my mom gave me permission, I quickly slipped on my shoes and ran there to find someone to play with me.

When I got to my aunt's house, I learned I'd just missed my sisters and cousins who had walked to the convenience store up the street to grab some snacks. As I started back home, I saw a teenage girl sitting on Aunt Grace's front porch. I knew her well. She was a foster child who my aunt looked after, sort of like a bonus cousin to me.

"Hi," I said, smiling innocently and waving from the sidewalk.

"Hey," she answered back, adding, "What you doing today?"

"Nothing," I answered. "Do you want to play?"

"Sure," she answered as she stood up. "Let's play inside," she suggested.

In my excitement and innocence, I nodded enthusiastically, forgetting the rule my aunt had laid down: *Never be alone with any of the foster children.*

She led me into her room and closed the door behind me. I don't remember all the details, but we played a game like Simon Says. In her game, though, most of what Simon said involved her touching various parts of my body.

I trusted her. To me, she was another member of the family. I went into her room where I thought we would play with toys; instead, she molested me, a concept far above the head of a four-year-old child. If that wasn't hard enough, after hearing what her foster daughter did to me, my aunt blamed me. She forced me to apologize to the girl. Then my aunt said that I could never step foot in her home again.

Although this experience traumatized me, my childhood had far more joyful moments than scarring ones.

For example, my cousins and I produced countless dance routines in our living room. As a child, I just knew one of us would go on to choreograph dances for Destiny's Child, B2K, or Britney Spears. Sadly, that never happened. Those dance routines I spent hours developing and mastering never translated into transferable adult skills off the dance floor. Just the same, I will two-step my way for as long as I live.

Not long after we got settled into Massillon, my mom delivered some unsettling news as I approached eight years old.

"I know we just moved into our new house, but I got a job offer. We're moving to Canton."

"Mom, I just started high school," my oldest sister, Alexis, protested. "I want to finish with my new friends!"

"I wish we could stay, too," my mom said, trying to ease the blow. "But I have to pay the bills. I don't have the gas money to make the commute to Canton every day. If y'all have a better idea, I'm all ears. But I can't leave you behind, Alexis. You're only fourteen."

"But I can just stay with Gran until I finish school," Alexis pleaded. "Then I'll move to Canton with you once I finish."

"Fine," Mom responded. "If my mom says you can stay with her, I'm fine with that. Now," Mom added, looking at Brandy and me, "I suppose you two want to stay with your grandmother, too?"

Brandy was eleven, and I was three years behind her. While we didn't want to move, we could not imagine being away from Mom.

Mom's eyes filled with tears as she realized that Brandy and I might also choose to stay behind with our Gran.

"No," I said quickly, seeing the pain in Mom's face. "I want to be with you! I'm excited to go to a new school and make new friends!"

My mom gave a subtle sigh of relief and offered a half-hearted smile. Brandy didn't protest the move, either.

What I said about being excited to start over with a new school and friends was a lie. The thought of moving devastated me. Not only would I be separated from that huge block of familiarity surrounding school life, but I'd also be leaving my extended family. And my immediate family, if Alexis stayed behind. I was heartbroken, but even worse, I could see that my mom was heartbroken, too.

For as long as I can remember, people have called me an old soul. I can't say I always knew what that meant or even if it was a compliment or insult. But as I've gotten older, I've come to understand that being an old soul means you feel things in your core that your brain might not fully comprehend, or your tongue might not be able to express.

Old souls see things—*know* things—that other people miss.

As a child, I watched my mom pile food on all of our dinner plates, except for her own. My mom always served herself half of what she served me. I saw that and took it in without saying a word. My sisters and I typically wasted a lot of food, usually pushing it around more than eating it. But then one day, I watched my mom clearing the plates from the table. Instead of scraping them immediately into the garbage, she ate the food we left behind. Mom made sure we were fed, even when it meant that she would do without or eat our leftovers.

Most people make random observations about what's going on around them. But old souls don't just take it in; they *do* something about it.

After I saw my mom eating from our plates, I always ate as little as possible, because I wanted to make sure that she had enough to eat.

Chapter 2

Potential in Search of Direction

"The only person you are destined to become is the person you decide to be."

~ Ralph Waldo Emerson

S INCE I WAS a kid, I always dreamt I would be something in life. Something big. People around me always told my mother, "She's going to be somebody! She's going to change the world, you just watch!"

Everyone saw this "potential" but could never put their finger on what exactly I had potential for. I didn't either. I had no general direction regarding where or what I wanted to be.

The first time someone told me that I was going to "do something special," I was twelve. My sister Alexis modeled for the retail chain Fashion Bug. Because of her beautiful skin, hair, and outgoing personality, she got invited to join a runway show to model their clothing. Since my sister couldn't drive, my mom took her to the event, and I tagged along to support her. As Alexis put on her first outfit for the day, I started dancing to the music playing in the background.

"Let's go," a lady said to me while I danced around like I owned the place.

"Go where?" I asked. "I'm not part of the show. I'm just here with my sister."

"Come here," the coordinator said. "You've got something very special," she told me while putting her hand on my chin to turn my face to different angles. "We could use you today, okay?"

Before I could answer, she started pulling clothes off a rack. "This one," she said as she handed me a blouse, "this will go nicely with this here." She finished by handing me a skirt. "Go," she snapped. "Get changed."

Before I knew it, I changed clothes and walked the runway, still strutting my dance moves like the carefree twelve-year-old I was. I stole the show!

After that event, I attended and graduated from Barbizon Modeling School. While I was there, they held a talent scouting event, and one of the judges was Josh Duhamel from the *Transformers* franchise, who was married to Fergie at the time. I had no idea who he was, but my sister Brandy did, and she stared in awe while I continued quietly going over my lines. When my turn came, I performed my script, and he loved it! He loved it so much that he awarded me his "golden ticket" to go to California for the International Modeling and Talent Association conference. Unfortunately, the "golden ticket" didn't come with the money to fly out to California to meet the talent scout.

"If they want to shell out money to take the two of us out there on their dime, we'd go," my mom explained. "But we're not going out there without a paycheck attached to it."

While I was disappointed, I shot a commercial and got paid for the use of my photos for advertising. After that, my modeling—and acting—career died before it got started. But for the time it lasted, I really did feel special! And I had fun doing it, too.

Another mother at the show told my mom, "Look at her spunky, outgoing personality. You just watch. I can't put my finger on it, but she's going to end up doing something special." That stuck with me.

In school, I didn't always feel special. Maybe that's because I didn't belong to one specific group. Instead, I played the drifter, fitting in with stoners and scholars, jocks and jerks, nerds and New Agers. I loved people, and I always had an outgoing personality. If someone looked grumpy, I took it upon myself to make them laugh. If someone sat alone in the cafeteria, I'd join them. I found reasons to smile all day long, exchanging pleasantries and small talk with anyone close enough to hear me.

Some of my classmates might remember me as "the funny girl with food in both hands!" Guilty as charged. I enjoyed eating. As far as funny, no one would confuse me with the class clown, but I did enjoy finding humor in situations.

But I couldn't always laugh my way into the hearts of others.

When I first started school in Canton, there were two Black students: me and a girl named Britney. Up until that year, Britney had been a novelty as the only Black student in a sea of White faces. She let me know in no uncertain terms that she could not stand me, and I'm guessing it had to do with me taking away her distinct characteristic.

The two of us waged a tit-for-tat war. She would steal stuff and put it in my locker to set me up, but she would do it while I stood right there! Then I would steal stuff and shove it in her locker. Things between us never got physical, and I was glad for that. As one of the smallest in the class, I didn't kid myself that a fistfight would go in my favor. But even though I didn't fight, I also didn't back down.

Once, four girls at school jumped me. I mustered up every bit of scrappiness I could find in my little body. If we had been in a boxing ring, the referee would have held up my hand and proclaimed me the winner. The only "proof" that I won is that I got suspended for three weeks. The other girls got suspended for one.

It wasn't long before my mom became *best friends* with the school principal. The two of them spent so many hours together, my friends may have thought they were dating. Their relationship brought me some sunshine as well as darkness. On the plus side, the principal listened to my mom and demonstrated fairness and impartiality. On the downside, other students thought my mom and the principal were all friendly, so they bullied me even more, saying, "You gonna tell your momma to tell the principal on us?"

I ended up earning the attention of another bully named Debbie. For two years, Debbie grabbed me in the cafeteria, dumped my tray, and stole my graham crackers. My crime? Having an item that she wanted.

Most of the time, I wore either gently worn hand-me-downs from my sisters or clothes from Kmart and Burlington Coat Factory. But occasionally, my mom would get me a nice name-brand outfit she found on sale. If I were lucky, I'd wear those outfits a few times before they were stolen from my gym locker by a couple of other girls. Again, I was guilty of having something they wanted.

I wasn't raised with the name-brand mentality, so I couldn't figure out why anyone would want my clothes badly enough to steal them! My mom taught me to be thankful for what I had, and on my own I learned to be thankful to my mom, who sacrificed and worked her butt off to provide for me.

As I got older, I better understood the nature of bullies and bad actors, those who took what wasn't theirs, disrespected what they didn't comprehend, and hurt others just because they could. In my heart, I sensed that those people that lashed out at me did so because they themselves were hurting. While that level of compassion kept me from hating others, it never diminished my pain.

You can take crap from people for only so long. Sometimes I did fight back. Of course, not physically. Why start a fight when it could leave scars on your face? Instead of getting physical, I got mouthy! My mouth and quick wit got me out of every fight I got pulled into.

Alexis and I were the family "rebels" in high school. But not Brandy. Brandy was perfect. Hell, she still is. Have you heard about or seen how the middle child, Jan, felt on *The Brady Bunch*? "Marcia, Marcia, Marcia! That's all I hear!" Well, in my family, even though Brandy was the middle child, she set the standard for goodness and smarts. Introverted Brandy never did anything wrong. She was too busy reading books in her room to get into trouble. She'd stay in her room for so long, we'd forget she lived with us or what she looked like!

"Mom!" I'd scream when Brandy finally popped out of her room for food. "Some girl broke into the house, and she's heading towards the kitchen!"

My sister Alexis rebelled in her own way. Colorism is an issue that is deeply rooted in the Black community, and my sister was constantly an unwilling casualty in their discrimination towards darker-toned women. Alexis has beautiful, dark skin, and in the Black community, lighter-skinned people are generally treated better than those of a darker shade. Sadly for Alexis, she didn't always fit in with a lot of people in Ohio, so she acted out. It took many years for her to recover from the serious emotional and psychological trauma of being bullied.

Since Alexis went to a different high school, I lived in the shadow of Brandy, who was three years ahead of me in school. When I had class with a teacher who remembered Brandy, I can't tell you how many times I heard, "Oh, Brandy was such a pleasure! I've never had such a smart, focused student!" Usually, within a couple of weeks, those same teachers would look at me with eyes asking, "So, are you adopted?" But what they'd say was something like, "You are absolutely *nothing* like your sister." They didn't mean it as a compliment.

One evening, my mom sat quietly at the dinner table.

"Mom, what's wrong? Are you mad?" I asked.

"Girl," my mom said, "you are my gray-hair baby. You see these?" She pulled her hair away from her head and showed me some strands of gray. "Every one of these gray hairs are from you. Brandy never gave me an ounce of trouble, but you're going to put me in an early grave!"

I loved my mom, but we had a rocky relationship in my teen years. Rocky but strong. If you are the mother of a teenage girl, you get this. It didn't help that our personalities were so similar. At times, I couldn't stand my mom. And then an hour later, I'd say, "Do you want to watch *Kojak*? I brought us snacks!"

Our favorite activity together was watching TV. We especially loved throwback shows like the original *Wonder Woman* with Lynda Carter, *In the Heat of the Night*, *Diagnosis: Murder*, and *Good Times*. While we were bonding over these shows, she told me that some of her best memories were when she watched these same programs in primetime with her mother back in the 1970s. And here we were, a generation later, repeating the same rituals. During these special times, Mom would tell me stories from her childhood, about what it was like growing up in a different era.

My fondest childhood memory is lying across the foot of my mother's bed while she shared emotional stories about the forces that shaped her into becoming the woman she grew into. A close second favorite memory is waking each Sunday morning to my mom's version of "Reveille" (a military bugle call typically used to get people into formation), only instead of a bugle, she blared gospel music or songs from Earth, Wind, and Fire.

One evening while my mom and I were watching TV, she decided to channel surf to see if we could find a show to add to our growing list of new favorites. After a few minutes, she stopped on a crime show. Even though we started the show midway, we became spellbound and finished watching it. We didn't learn the name of the show until the program ended, and then we cheered as the next episode began immediately. The show was called *Criminal Minds*. We were instantly hooked and stayed up longer than we should have. The show was about a group of FBI profilers who studied the minds of criminals to determine their behavioral patterns so they could track and apprehend them. Not only did they bring about justice, but they also helped victims along the way.

"I can SO see myself doing that job, Mom!" I said at the end of the second episode. "Can't you see me catching bad guys and helping victims?" I asked rhetorically.

"Mmh-hmm," Mom nodded in agreement. "You can do anything you want to do when you put your mind to it."

In that moment, I figured out where I wanted to apply my potential. But I had yet to figure out how to get there.

Chapter 3

The High Cost of Free Stuff

"I may be running out of options, but running out isn't an option."

~ Mark Lawrence

A S MY MOM and I talked over the closing credits of *Criminal Minds*, I started thinking, *Maybe this is my something special. Maybe I can change the world by being a criminologist!*

The next day at school, I applied for the career tech program called Legal Studies. To get into the program, I would need to have grades at or above a certain GPA (I did by a hair!), and I had to be approved by the instructor.

"Okay, I know it's just a TV show," I told the Legal Studies instructor in our one-on-one interview, "But I found my calling!" I explained that I wanted to do something with my life that combined psychology and criminal justice. Criminology seemed the perfect fit.

Seeing my passion, the instructor accepted me into the program. And, armed with a purpose, I went from an average student to one who excelled in basic legal proceedings, business math, forensics science, legal terminology, and even mock trials. This program seemed custom-made for me, and with my goals in sight, I came alive.

As my junior year ended, my friends and I made a pact that we would live our lives to the fullest that upcoming summer before starting college early together. Since this would be my last summer to live before college and then adulting full-time, I planned to make it memorable. That's when I met my first crush. Yes! This would be a summer to remember.

This boy really liked me. Knowing nothing about playing hard-to-get, I figured if he liked me, I'd like him back. Enjoying my summer love, I floated somewhere near cloud nine. But I had a good head on my shoulders, which kept my feet grounded.

My mom didn't need to worry about me "getting into trouble" with a boy. I had my own standards, and I wasn't about to cross them, even with someone I loved.

A few weeks into dating, he began pressuring me to sleep with him. He didn't want to take *no* for an answer.

"Come on, Kia," he begged. "You know I love you. It's no big thing, and if you love me, you should let me love *all* of you."

I'll bet Adam used that same line on Eve, but I wasn't buying it.

"No," I said for what felt like a hundred times. "And I wish you'd stop asking," I pleaded with him. "I'm not going to have sex with you."

I never changed my answer. Eventually, he two-timed me, calling me his girlfriend while he slept with other girls. That's when I'd had enough and broke up with him.

Sometimes we suffer the consequences of our own actions; other times, we become victims of the bad actions of others.

While babysitting my nephew at my sister's house, my ex-boyfriend texted me.

"Leave me alone," I texted back. "I told you I don't ever want to talk to you again."

Unnerved by his text, I grabbed my nephew and headed out the backdoor to return to my own house, where I felt safer.

My ex waited for me in the shadows. After pulling a gun on me, he forced me back inside the house. Then, after locking my nephew in his room, he beat me until I started to lose consciousness. Then he raped me.

When I had been molested at four, I'd felt confused by what happened—and terribly ashamed about upsetting Aunt Grace. This shame, though, cut much deeper. I didn't want anyone to know what happened to me. But I had no choice but to tell Mom most of what happened. I looked like I'd been hit by a truck. Mom took me to the hospital, where I denied that I'd been sexually assaulted. I had so much shame I couldn't admit it to myself, much less my mother. They conducted no rape kit.

After being released from the hospital, my mom took me to the police station to report the physical assault.

"No one hurts my child," Mom said as she drove.

"Do you have any proof this happened?" the police officer asked after I told him about the beating.

"Proof?" my mom asked. "Look at her. Do you think she'd beat herself up like that?"

"No, not at all," the policeman responded. "But we have to have proof that this boy," he called him by name, "is the one who did this to you."

"No," I said, my tears splatting on the floor around my feet.

"Her nephew was there," my mom added. "He saw him."

"Did your nephew recognize him?" the officer asked me as he wrote a note on his pad.

"We asked him," my mom answered for me. "He kept saying, 'A man came into the house.' That's all he could say. He's four."

"It's too bad the boy is so young," the police officer said gently. "Even if he could identify the one who did this to you, I can't go to the district attorney with a four-year-old as an eyewitness. Children that age aren't considered reliable."

The police interviewed my attacker, who provided an alibi. He told them he was at church with his family. Church! His sister, of course, backed up his story. In the end, we dropped the charges against him. He got away with what he did to me. I'd been assaulted, but the rest of the world continued like nothing had happened.

But not me. This event gave me nightmares for years. And for just as many years, I tried to push those memories to the back of my mind. I never talked about it again, and I never sought counseling to help me process that attack. But from that moment on, I wanted to get as far away from Canton as I could.

Being sexually assaulted robbed me of more than my physical well-being. It created fear and anxiety I'd never had before. My thoughts teetered between processing the attack and pushing it out of my mind. In hindsight, by stuffing my emotions and pretending that everything was okay, I grew too distracted to act on building my future.

By the beginning of my senior year, many of my friends received acceptance letters from colleges. Not me. I didn't even have the right frame of mind to apply to colleges. It took all of my effort to do well in my classes and keep my fears at bay.

So when all of my friends were like, "I scored a blah blah blah on my SAT" and "According to Sallie Mae…" I felt like an utter, complete failure. I'd dreamed about my future, but I didn't have the emotional capacity to plan for it.

"I thought my biggest problem would be getting accepted into my first choice of college," my friend complained as we ran up the stairs to the cafeteria. "But that was the easy part. Now they want me to declare my major! I have no idea what I want to study."

"I'll trade you problems," I mumbled.

I had spent the last two years in the legal studies program. If I got into college immediately after graduation, I would earn my associate's degree, and my credits would transfer into an accelerated bachelor's program. But now I felt gutted, like I had nothing to look forward to. I had worked hard to pursue this dream of studying criminology, but I never turned that dream into a plan.

As my friend and I got to the top of the stairs, we saw some students standing near the door to the cafeteria. A recruiter for the United States Air Force hurriedly set up a colorful display with their insignia, flags, and brochures. Our school often hosted recruiters from the US Army, Navy, and Marines, but this was the first time I had seen the Air Force there. And then one more thing caught my eye: an Air Force backpack.

I should probably mention that right after my love of eating is my love for free stuff. I'm not proud of this, but I'm pretty sure I'd have been willing to walk over broken glass to get my hands on anything free!

The recruiter must have seen the gleam in my eye, because he stopped straightening his brochures and came up to me.

"Hi. What's your name?" he asked.

"Kia," I answered, hardly taking my eyes from the backpack.

"You like that backpack," he said by way of statement instead of a question. "How many push-ups can you do?"

"Depends," I replied with a little cockiness.

"It depends on what?" he asked with a smile.

"It depends on what you mean by 'push-up,'" I answered, and we both laughed.

Truth be told, I was pulling in a high (possibly low) C in physical education. PE was my lowest grade. I guess I didn't dodge the ball right or climb the rope fast enough.

The recruiter got down on the ground in front of me and showed me how to do a proper push-up. When he finished demonstrating the right form, he stood back up.

"Do eighteen push-ups, and this is yours," he said, holding the backpack in the air. "I'll even throw in an Air Force water bottle."

"No problem," I said, eager to win this contest.

"But before you do those push-ups, I'll need you to fill one of these out," he said, handing me a clipboard with a form and pen attached.

Moments later, he counted off as I cracked off my pushups.

"Twoooooo," he said, elongating the word to reflect how long it took my arms to lift my tiny body into a push-up.

After what felt like hours, I was certain I must be nearing the goal.

"Niiiiiiiiiiinnnnnnnnnne," he stretched out his count while my arms trembled and sweat dripped off my face.

You gotta be shittin' me! I thought to myself. *Only nine?!*

Long story short, I managed fourteen asthmatic pushups before my arms gave out.

I failed again, my own voice echoed inside my head.

I closed my eyes and remained on the floor. Sweat from my effort mingled with the hot tears of my frustration and sense of hopelessness. I knew that I had the brains and the discipline for college, but I had failed to apply in a timely manner. And I'd had no assistance from the guidance counselor.

"Hey," the recruiter snapped me out of my self-pity. "Here," he said, offering me his hand to get off the floor.

"Thanks," I managed to say.

Patting me on the back, he added, "If you can do fourteen pushups without any practice, you'd have no problem doing eighteen if you worked at it."

"Thanks," I said again, still disgusted with myself.

"Don't ever give up," he said as he threw me the backpack with a water bottle stuck in a side mesh pocket. "I'll be in touch soon."

That weekend in my room while I started to Ask Jeeves, Yahoo, and Google to find a college that would let me in at this late date, my mom screamed from downstairs.

"Jakia, get your ass down here!"

She's even mad at me when I'm just in my room doing nothing! I thought to myself.

"What is it, Mom? I'm looking at colleges," I explained at the top of my voice.

"I said *get your ass down here,*" she repeated with a little more growl in her voice.

Since I wasn't sure if I had done something wrong to earn Mom's anger, I rushed downstairs.

"What's going on?"

"What's going on?" she mimicked me sarcastically. "You want to tell me about Sergeant Peters?"

"Who?" I shook my head. "I have no idea what you're talking about."

"He's from the military," my mom spat out. "How many times I got to tell you there's nothing free in life? I see that new backpack in your room. You signed a damned contract with the military to get that bag."

"No, Mom, that's not what happened," I sputtered. "I just had to do push-ups…" I started to say.

"He said you signed a contract, and he's coming over to see you at two o'clock this afternoon," she told me as I freaked out.

For the next two hours, I paced while waiting for S.Sgt. Peters. I couldn't focus on looking for schools. I couldn't even eat, which had never happened before in my life.

Goodbye, dreams of college, a voice spoke in my head. *You signed a contract with the military.*

S.Sgt. Peters arrived promptly at 2 p.m., and I invited him inside to the living room.

"Listen," I said as soon as we were seated, "Sergeant Peters, I didn't know I signed a contract with you. I just wanted the backpack, and…" Tears started building in my eyes.

In hindsight, *contract* sounds a lot like *contact*, don't you think?

I ended up enjoying talking with S.Sgt. Peters, so when he offered to take me to the recruiting station to tell me more, I got excited.

A career in the military piqued my curiosity. Coming from Canton, most of the people around me were passionate about sports and dreamed of making it professionally. Yet most of them ended up in blue-collar fields once that dream died. The military was something different.

"You don't have any firm college plans yet?" S.Sgt. Peters asked me when I joined him at his office in town.

"Nothing concrete," I said, while I could have added, *Nothing in Jell-O, either.*

"That's great," he said enthusiastically. "Why foot the bill or go into debt for college? If you join the Air Force, Uncle Sam will pay for your schooling while you serve our country."

It didn't take me long to weigh my options. Because I didn't really have any viable options. I still wanted to go to college, but I couldn't attend with my friends as they started school in the summer. I knew I would eventually figure it all out. My choices boiled down to joining the military as a way of getting out of Canton or becoming one with the couch in the living room until my mom threw me out of the house.

"You were interested in criminology, right?" he asked.

"Yes, sir," I answered smartly. "That's what I want to do with my life."

"And the Air Force can help you with that," he said cheerfully. "You can start working on your law degree while serving."

Manipulation requires two active partners. The manipulated party must be gullible and trusting. Getting manipulated doesn't make someone stupid; it simply means they are trusting and desire to see the best in others.

Manipulators rely on partial truths to sell their lies. S.Sgt. Peters offered me the stars and the moon to join the Air Force. Partial truth. There really are stars and a moon, but they weren't his to give me.

It's easy to fall for a lie when you want it to be true.

Have you heard the story in the Old Testament of the Bible about Esau trading his birthright to Jacob in exchange for some porridge? Little did I know I had just sold mine for a backpack, a plastic bottle, and a pretty little lie.

Chapter 4

Bulking Up

"The value of achievement lies in the achieving."

~ Albert Einstein

B EFORE SEEING THAT Air Force backpack, the idea of joining the Air Force never had entered my mind. In fact, I don't know if I knew the United States had a branch called the Air Force. Had I been asked to name the different military branches, I would have said Army, Marines, Navy, and Coast Guard. And unless you happen to be in the Coast Guard, you probably wouldn't consider them a "real" military branch. But Air Force? Nope. It never registered.

But as the thought of entering the Air Force seemed possible, my eyes were opened. Watching TV one evening with my mom, an Air Force commercial came on, and for the first time, it clicked.

"Mom, look!" I shouted with excitement. "You see that?" The commercial showed beautiful, young, smart-looking people flying and working on jets, all of them happy and wearing confident smiles.

"That's going to be me!" I exploded.

For the first time in a while, I allowed myself to feel good about my future. I had high hopes. I would be joining this elite part of the government, defending my country, living with a sense of purpose, finding myself, seeing the world, and getting my education. As a bonus, my mom would no longer have to worry about me. I had watched her scrimp and save to help send Brandy to college, and I didn't want to add to her burden.

But there was more to it. Joining the military would get me out of Canton. It would ensure I'd never again see the boy who assaulted me. By moving away, doing something meaningful and positive, I believed I could leave the past in the past.

Maybe that's why I thought, at just age seventeen, that the world fell completely wide open to me! Why wouldn't I want to join the Air Force? But, it turned out, wanting to get in was easier than actually getting accepted. I still had some hoops to jump through before I could be called an airman.

First off, I had to overcome my mom's objections. I knew better than to go at it alone with my mom, so I brought S.Sgt. Peters with me.

"I was born at night, but not last night!" she spat out at me. "What are you going to do when you're deployed to a war zone?" she asked. "Did you think about that? Cuz I did!"

"Ma'am," S.Sgt. Peters broke in before I could respond. "I can tell you that since she's a female, she's much less likely to be deployed than if she were a male. So that's not really something you need to be worried about."

That turned out to be a lie, but we didn't know that at the time. Just the same, it worked on my mom.

A few days later, S.Sgt. Peters brought over the official contract, that one that set the wheels in motion.

"I don't want you to sign this unless you are both 100 percent comfortable with what it says," S.Sgt. Peters told us, helping sell the idea further to my mom. "And since Jakia is still seventeen, she can't join the Air Force without your support and consent on this contract."

Then he went through the contract, explaining every detail. He began with what I could expect training to look like, what kind of job and experience I could gain, and the military benefits including healthcare and college. By the time he finished, I think my mom wanted to join, too!

"I know she will be in good hands," my mom said reluctantly. "But she's my baby. We've never lived apart, and I don't know that I can have her living on the other side of the country."

"I hear that all the time," S.Sgt. Peters said soothingly. In hindsight, I'm guessing that recruiters memorize a script with pat answers to address any pushback they might encounter. "But this is a new day. No matter where she goes in the Air Force, you can Skype and use other technology that makes it like you two are in the same room."

I wanted out of Dodge—or Canton—and this seemed like the best path for me, one that made me excited about my future. And I also wanted to make my mom proud of me. Too many times in my life I had worried my mom or put gray hairs on her head. I felt like this move would help me figure out who I was—and get out from under the shadow of my sister, Brandy.

As part of the signing process, S.Sgt. Peters reviewed various jobs I could work in the Air Force, jobs based on my scores on the Armed Services Vocational Aptitude Battery (ASVAB) that I had taken a few weeks earlier. Since I scored well, I had many options open to pursue. I chose to pursue a job where I could help others: diagnostic imaging. My mom worked in healthcare, so I knew those jobs were in high demand.

I figured studying diagnostic imaging would give me job security after I left the Air Force.

"According to the ASVAB results," S.Sgt. Peters said as I entered diagnostic imaging onto the form, "I'm sure they will fast-track you into that program. You're a shoo-in!"

That also turned out to be a lie. In fact, S.Sgt. Peters recruited several people from Canton, and he told similar lies to many of us. He promised one friend of mine that she would end up as a paralegal; she ended up flipping burgers in the dining facility (also known as DFAC). One time we all joked that if we ever met S.Sgt. Peters on the street, we'd kick his ass.

"Remember me?" I could imagine myself saying to him as I sized him up.

When Mom finally signed the document and handed the clipboard with her signature on it back to S.Sgt. Peters, I got choked up

"Now," he announced, "there's just one more thing I need to let you know. The Air Force has weight minimums by gender and height. According to your application, you are 5'5" and weigh ninety-four pounds. Is that right?" he asked me.

"That's about right," I said.

"Okay, we are going to need to work on that. This height and weight chart tells me that you must weight no less than 105 pounds to join the Air Force. Do you like to eat?" he asked.

Music to my ears!

"I love to eat!" I yelled, feeling like I'd won a second lottery.

In a perfect world, I could have become a competitive eater. Sponsors would pay me to eat their food, and event coordinators would charge people admission to watch me stuff my face. Had I chosen that career path at age seventeen, I'm sure I would have been world-class by now.

But it turned out that as much as I loved to eat, I could not for the life of me gain weight.

That's when my little village came together to help me. The man my mom had dated for years had a plan. As a child, I called my mom's boyfriend "Here Daddy" since he played such a strong positive role in my life. I referred to my biological father as "There Daddy" since he lived in another state.

"What do you mean my baby might not be accepted into the military?" Here Daddy asked in astonishment. "No, we're not going to have that. If there's one thing I know how to do, it's gain weight!" he said enthusiastically.

Here Daddy's weight-gaining program consisted of whey protein, pizza, ice cream, and any other food in sight. As soon as I woke up, I ate. Before going to bed, I ate. In fact, I stopped sleeping in my bed upstairs. Afraid that exercise would burn precious calories, I slept downstairs on the living room couch to avoid taking extra steps.

"Here Daddy!" I shouted into the phone after I finished a huge container of whey protein in a couple of days. "I gained five pounds!"

An hour later, I peed. And lost five pounds.

In January, S.Sgt. Peters sent me to the Military Entrance Processing Station (MEPS) in Cleveland, Ohio, to see what the process was like. After a full month of sitting still and eating until I felt like I was going to pop, I weighed ninety-eight pounds. But they didn't reject me outright. I considered that progress!

After I "failed" weigh in, I called S.Sgt. Peters, who developed a workout program for me. Part of his program included a backpack full of high-calorie snacks.

"You know about the carrot and the stick for moving a donkey?" he asked. "Well, these donuts here are like carrots."

We started running together, him in front of me, usually running backwards, taunting me with the goodies I knew he had in his pack.

"You want these donuts I got back here?" he yelled from up ahead of me. "You better come get them before I eat them all!"

That would make me run faster. We made a game of it. If I could catch him before he got to a certain landmark, we'd stop, and I could have whatever I wanted out of the backpack of treats. And if I beat him, he'd give me a gift card to my favorite restaurant! The best part was that my new weight gain wasn't fat. It was solid muscle.

The night before my final visit to MEPS, S.Sgt. Peters called me.

"I need you to eat everything you can fit in your stomach, and drink water like your future depends on it!" he said. "Because it does, Jakia. I know you want this. And I know you can do this. You're motivated. You've got this. Now don't show me with your words. Show me on the scale!"

The next morning when I got on the scale, I took in a deep breath as if the extra air would add an ounce or two. The needle bobbled around a bit before it rested on the number: 106. I weighed 106 pounds! I beat the minimum by a whole pound!

"I did it, Mom!" I screamed across the house. "I did it! I'm 106 pounds! I'm going into the Air Force!"

Before heading to MEPS, my mom drove me to meet with S.Sgt. Peters one last time. I took my bags with me. We said a final, tearful goodbye at the recruitment station. My sister Alexis wanted to come, but her boss wouldn't let her off work. My sister Brandy was still away at college, so she couldn't see me off, either. As a nice gesture, S.Sgt. Peters arranged a chauffeur to drive me to MEPS in Cleveland.

The warm Indian summer air embraced my cheeks outside the doors of the processing station. As soon as I got inside, I didn't need to worry about getting lost. Someone kept yelling instructions to keep us in line and moving.

"Females, line up outside the women's restroom to the right. Males, line up outside the men's room on the left," a sergeant announced loudly as dozens of recruits milled around inside the building.

Taking my place in line, I looked at the girls around me. I felt like a kid surrounded by giants. *But that's okay,* I calmed myself. *It's not a competition. I made weight. As long as I don't have to pee, I'm gonna make it.*

So, of course, the first thing the sergeant says as we enter the restroom is something like, "Females, you'll be given a cup. When it's your turn, you'll need to fill it."

Damn, I thought.

"You're next," the tester said as she pointed me to a toilet. I walked into the doorless stall and turned around.

"Where's the door?" I asked in shock.

"You don't need a door. Just fill the cup," she responded.

She might not have needed a door, but I felt extremely uncomfortable without one.

"Okay," I said, trying to look confident. "Well, could you at least stand aside to give me a little privacy?"

"No," she answered flatly.

The woman stood right there. I tried shifting a little to the left, to give myself just an iota of dignity.

"Stop moving," the woman said sternly. "Just spread your legs and go."

What the hell kinda place is this? I started wondering.

I hoped to relieve myself just enough to fill the cup and then stop, wanting to hold as much in as possible. I tried to keep everything tight, to fill the bottle a drop at a time. Yeah, that didn't work. Once I got started, the whole dam burst.

And there went all my weight, I told myself, fearing that I had just lost my final chance of getting into the Air Force. I handed her my sample, washed my hands, and went into the larger room to wait with the others.

This open room looked like a gymnasium with no windows. It had flooring like a basketball court. We females stood in our bras and underwear trying to look casual, like this was just a typical day for us all. But I was not looking casual. Instead, I looked nervous and worried.

"What's wrong, girl?" a couple of girls asked me while the others were still providing samples in the bathroom.

"I killed myself trying to make weight," I said shakily. "But I'm afraid I peed out so much that I'm underweight now. I don't want them to send me home again for not making weight!"

"Girl," one recruit said in a hushed tone, "there's a sink right over there. Go ahead and drink as much water as you can." While one girl helped me to the sink, another girl kept lookout for the sergeant.

I swallowed several mouthfuls before a booming voice told me I should quit drinking and get back in formation.

"Time to check your height and weight," the sergeant bellowed in the large room, her voice echoing off the walls.

They lined us up again (lines, I would learn quickly, were a big part of military training and life) next to a row of lockers next to a scale.

"Next!" the sergeant yelled without looking up from the scale. Yelling, too, I would learn quickly, was a big part of military training and life.

Finally, my turn came.

"Next!" she yelled.

As the sergeant bent down slightly to better read the numbers on the scale, I placed a hand on the lockers and pushed myself down as hard as I could to add any extra weight possible.

The numbers of the scale bounced all around as I struggled to keep steady pressure on the lockers.

"Stand still!" she yelled without looking up.

"Sorry," I said meekly. "I'm just nervous."

Finally, the digital scale locked out and blinked once: 105. I made it!

Then they checked our heights. My recruiter explained that BMI factored into getting accepted. Since I measured 5'5", S.Sgt. Peters told me to try to bend my knees a little to lower my BMI to height ratio. Following his instructions, I bent my knees.

"Stand up straight," she yelled, noticing my knees.

Then I tried standing like the Hunchback of Notre Dame.

"Do know what the word straight means?" she asked sarcastically.

Somehow, I managed to look like I stood up straight while twisting like I had scoliosis. By the time they finished recording my measurements, I weighed 105 pounds and stood 5'2" tall!

I'm not proud of myself, but I did what I had to do. When testing was complete, we all cheered. The girls who had helped me gave me hugs and pats on the back. I felt like I had joined an incredibly special sorority.

After all recruits were weighed and measured, they had us get dressed again and brought us into a large room, both males and females.

"Congratulations, recruits!" a sergeant yelled while we stood at something resembling attention. "Before you leave here and report to Basic Military Training, please raise your right hand and repeat after me the United States Armed Forces oath of enlistment."

Across the large room, recruits straightened themselves and right hands shot up in waves.

"I, (state your name), do solemnly swear," the sergeant began.

"I, Jakia Monee Clark, do solemnly swear," I repeated as my face flushed hot with pride and sudden emotion. "That I will support and defend the Constitution of the United States against all enemies, foreign and domestic; that I will bear true faith and allegiance to the same; and that I will obey the orders of the President of the United States and the orders of the officers appointed over me, according to regulations and the Uniform Code of Military Justice. So help me God."

By the time I finished, tears poured down my face as the realization washed over me that I now was part of something bigger than myself. I had fought hard to be here in this room and to join such an elite group of military personnel. No matter the missteps that had brought me here, I was now a future airman. Pride beamed from my face.

Instead of returning home that evening, I stayed in a hotel room. I called S.Sgt. Peters to tell him the good news. Then I called my mom, and it was as if everything had changed between us.

"Baby, you are going places," my mom told me over the phone.

I didn't know it then, but the abuse I'd endured in my childhood and teen years would follow me as I took my first steps from adolescence into adulthood.

Chapter 5

Trainee Dumb-Ass

"I'd rather have names to hurt me, than my bones broken with sticks and stones."

~ Anthony Liccione

DAYS AFTER OFFICIALLY entering the Air Force, I flew to San Antonio, Texas, to begin eight and a half weeks of Basic Military Training. To say that I left for Basic with few details would be an understatement. S.Sgt. Peters hadn't prepared me at all. He gave me just two pieces of advice, the first one being, "It's a long bus ride to base, so wear something comfortable. It doesn't matter what clothes you bring, because you're going to get free clothes as soon as you get there."

I wish he had been more specific. I wore some broken-in, ripped jeans and a T-shirt with cutouts in the back. They were comfortable, and I thought I looked nice.

As the bus pulled away, I realized that I wasn't in Kansas—or Canton—anymore.

I remember a lot of yelling. For the life of me, I couldn't figure out why anyone would be yelling. *Why are you so angry first thing in the morning?* I wondered as the sergeant on the bus kept shouting and saying things that made absolutely no sense to me. When I entered the bus, I was full of happiness verging on joy. But my cheerfulness waned once I realized I was on a long road trip with a loud and somewhat hostile man.

As the bus rolled up to the base in San Antonio, I figured we looked like a scene out of *Shawshank Redemption*. But instead of the existing "prisoners" sitting around placing bets on which of us newbies would die first, airmen-in-training avoided eye contact with us. They were too busy marching, running, doing calisthenics, and getting yelled at to take much notice of us.

I snapped out of my thoughts when the sergeant screamed from the front of the bus, "Which of you dipshits play a musical instrument?"

Very quickly, I remembered the second piece of advice S.Sgt. Peters gave me before I left Canton.

"You might be tempted, but don't volunteer for anything," he said. "Don't raise your hand. Trust me on this one. Just don't do it."

S.Sgt. Peters hadn't been right yet, but to be safe, when the sergeant asked for volunteers, I kept quiet even though I had played the clarinet in school.

A few others must have missed Peters's memo, and they cheerily raised their hands.

"If your hand is up, get the fuck off my bus," he yelled in gratitude for their volunteering spirits. "You are going to *band flight*."

In the Air Force, a flight refers to two or more airmen or sections of airmen. In this case, those unlucky volunteers would join the band flight, meaning in addition to going through Basic like the rest of us, they had to play an instrument—and carry them around the entire time.

For the first and only time, I appreciated S.Sgt. Peters for helping me dodge that bullet.

"Any of you dipshits have any aches or pains on your body?" the sergeant asked the rest of us who remained on the bus.

I had a leg cramp, but I figured it hurt a lot less than what might happen to me should I raise my hand. I kept my trap shut.

After a few "dipshit whiners" raised their hands and were ordered off the bus, the rest of us were herded off *en masse* in front of a large building.

"Move! Move! Move!" the sergeant yelled. "Hurry up! MOVE, MOVE, MOVE!" A moment later, he added to the confusion when he said, "I said, 'Hurry,' not, 'Run!' Move your fat, lazy, slow asses!" He continued shouting as we lined up outside what I assumed were our future barracks for the duration of boot camp. Adding to these mixed-messages was the confusion of other angry-sounding leaders yelling different things from different parts of the yard.

After I stepped out of the bus, a female drill sergeant stopped in front of me while we stood in formation.

"You dressed like a whore to come to Basic?" she asked, her face wearing an expression of utter disbelief.

Ironically, those would be the nicest words I would hear for a while.

Had I known anything about the Air Force, I would have known that they have *military training instructors*, not drill sergeants. And we were called *trainees*. We had not yet earned the right to the title of airman, which comes after completing Basic Military Training, not boot camp. And the Air Force doesn't have barracks; they have *dormitories.* But what did I know? I hadn't even watched *Top Gun*! Had I seen *Full Metal Jacket, Platoon,* or even *Stripes,* I would have had at least a small clue about what to expect. Instead, I went in cold.

"Look at you," the military training instructor continued after saying I'd dressed like a whore. "You're dressed like Trainee Dumb-Ass." That little pet-name stuck and became my new name throughout Basic. It hurt at first, until I found myself in the company of Trainee Fat-Ass, Trainee Stinky-Ass, Trainee Ass-Kisser, and Trainee Lazy-Ass. I felt like I got off easy.

After lining up, another female training instructor took us inside the dorm. And that's when things got a little weird. She assigned all of us bunks and wall lockers.

"Touch your locker," she instructed, so we did. "Now touch your bed," she said, and again, we did as she asked. "Now touch your locker, touch your bed, and touch your locker again," she ordered, and we did as we were told.

"These things are yours," she said as if we were a kindergarten class. "Now, take off your clothes!" she screamed. "Do it, do it do it now!" she bellowed as we stood there with our eyes open wide, flinging off our clothes, more afraid of her than the awkwardness of stripping in a room of strangers.

Once naked, she said something that made my mind lock up for a moment.

"Now take your body wash and squirt it all over yourselves!" she yelled. Remember, I was just a kid and not very worldly. But this started to remind me of what a girl in high school told me she had seen in a porno.

"Take your shampoo! Put it on your heads!" she screamed as we hesitated only briefly. The whole time, she just stood there staring at us. A couple of trainees closed their eyes, and more than a few shot sideways glances at other trainees with a look that clearly conveyed, "Any of y'all think this shit's a little crazy?"

"Now get your stinky asses in the shower and clean up!" she yelled. We were more than happy to go anywhere she wouldn't be.

Ironically, this event led me to meet one of my best friends, Kara. After the weirdness of those few minutes, we were all a bit tense. But Kara knew how to relieve the tension. All crammed in the shower, she looked over her shoulder at me and said, "Hey, you wanna hit my back?"

That broke the ice. Several of us laughed, and I jumped in, "Sure, I got you!" Then I turned around to Kara and asked, "Hey, can you get mine?"

Those few moments of goofiness started our friendship. From Basic, I keep in touch with eight friends. We stay connected on social media and through email.

Funny, when I arrived at Basic, they took measurements again.

"How the hell did you lose ten pounds and grow three inches since last week?" the medic asked as he recorded my measurements.

"Growth spurt, sir?" I answered.

Due to my low weight, they gave me an eating waiver that meant I got to eat at a special table and consume double rations for the duration of Basic. As a bonus, instructors could not yell at us or do anything to induce stress while we were eating. We were treated with kid gloves, if only at mealtime. Meanwhile, all others in training had to scarf down their meals while getting yelled at like they were eating under attack.

On a typical day at Basic, we would awaken to the sound of "Reveille" at 4 a.m. while the sky remained black. Gone were the days of sleeping until the last minute before catching the bus for school. I learned that as soon as my eyes popped open, I had to move quickly to throw on clothes and stand beside my bed. After a few days, I didn't even think about a snooze button option, and I stopped trying to figure out the day of the week or why the hell I was waking up to a shrill bugle in the middle of the night.

Once clothed, several of us stood at attention next to our beds with our shirts on backwards or inside out and socks missing. But we knew the drill. We had to stand there waiting for our military training instructor to enter the dormitory. Our primary training instructor was male, meaning he couldn't enter the dorm while we slept. But sometimes for fun—theirs, not ours—the sergeant would send in a female training instructor in the middle of the night. More than once, I woke out of sound sleep to see a face an inch from my own.

"Get your dumb ass back to sleep!" she would yell into my nostrils.

I got used to it after a while. My older sisters used to play mind games on me. Nothing quite this extreme, of course. I understood these tricks were an attempt to break us. If we broke in training, we would be unprepared for combat.

After the instructor checked out our flight, he ordered us downstairs to form up where he again sized up our flight. Next, he'd take us to the drill pad, where we'd do morning exercises until the sun came up. We'd do push-ups, sit-ups, flutter kicks, crunches, pull-ups, and run our butts off while praying the sun would pop up over the horizon so we could rest.

Let me put these brutal mornings into context. Most of us had just finished high school, putting us at seventeen or eighteen years old. Think about what you were doing at that age. Many non-military teenagers are just coming in from a night of partying or hanging out with friends at 4 a.m. But there we were rousted from sleep, thrown together, marched outside, and performing strenuous exercise before we would ever be out of bed back in our civilian days—which seemed so far away from this world.

After we were dismissed from exercise, we would rush upstairs and jump into the group showers. This was no place for modesty. Back in high school, most girls were too ashamed of their bodies to shower after gym class even when it meant going into semi-private shower stalls. But in Basic, we were too tired to feel shame and too afraid to be late to have any other concerns.

After showering, we'd put on our uniforms and make our beds. I learned quickly that what passed for "making my bed" at home didn't come close by Air Force standards. I wondered, *How many wars are won by a well-made bed?* But I kept my sarcastic thoughts to myself. I knew that making the bed the "right way" was all about discipline, taking criticism, and learning to conform to a consistent way of doing things. In fact, it didn't matter if our beds looked perfect. The instructor would typically toss each bed, throwing mattresses and sheets all over the bay, while yelling something like, "If you cannot even make a bed correctly, how in the hell is this fine military institution going to trust you with something important, like being out on deployment when lives are at stake?"

After completing inspection, we came to my first favorite time of day: breakfast. I loved the food at Basic. In hindsight, I don't know if they served us anything out of the ordinary, or if my hunger grew exponentially with all the physical activity. Either way, I usually finished a couple of plates of whatever they served up.

After breakfast, we'd form up outside before cleaning the dormitory and, of course, re-making our beds. Then, just like at home, it was chore time. And, just like home, the Air Force didn't have any "glamorous" chore options. The laundry people did the laundry. The bathroom people cleaned the bathroom. Some people spent the entire time lining up bunks, trunks, and shoes so straight those items could have served as a plumb line to lay out a pyramid.

Next, we'd attend school, learning Air Force 101. In the classroom, we learned military history, what it means to be in the military, the rank structure as well as customs and courtesies. The goal of these classes was to teach us how to think and speak like the military instead of civilians.

After class, lunchtime! Afterwards, we headed back for a few more hours of classroom instruction. Just like in school, we were tested on everything we'd learn.

After class, we'd do drill. Think of drill like football or cheerleading practice. When you watch athletes perform during a competition, you're watching hours of training that went into creating precise, graceful movements. If a team hasn't drilled, it shows. If a team has drilled, you see them functioning as one—instead of a ragtag collection of individuals. In the Air Force, drill covers all movements that will be on display at graduation. That means hours of marching, lining up, standing in formation, and working together as a team. It's all designed to instill basic, necessary discipline and uniformity.

After drills, we'd head back to the dorm to finish any chores left incomplete from the morning. At random times in our dorm, the military training instructor (MTI) would play music. Whenever he played the song "Let the Bodies Hit the Floor" by

Drowning Pool, we had to stop what we were doing, drop to the floor, and execute a series of push-ups, flutter kicks, and sit-ups until the song ended.

At the same time we finished our chores, the MTI held amnesty hour and mail call. Now that I've seen a few military movies, I recognize this concept from scenes where someone of lower ranks asks a member of senior rank this question: "Sir, Trainee (insert name here) reports as ordered. Permission to speak freely, sir?"

Before initiating any conversation with a MTI in training, we had to use what is called a reporting statement. That meant you didn't just start talking or asking questions. Instead, I would have to say, "Sir, Trainee Clark reports as ordered." To do otherwise would incur the wrath of the MTI, which if I got lucky would just be a verbal lashing like, "Trainee, did you just dare speak to me without a reporting statement?!"

But when you said the magic reporting statement, you could say pretty much whatever was on your mind to your MTI during amnesty hour.

"Why you always coming down so hard on me?" I asked during one of my amnesty visits. Another time, I made my complaint more personal like, "The Air Force must not want thinking trainees, because every time I make a suggestion, you say, 'What the piss, trainee?'"

During amnesty hour, we were encouraged to say whatever was on our minds without fear of retribution. The training instructor would just listen. And sometimes, they'd get real with us, and we'd get a sense that somehow, they, too, were human.

"Look, I've been in the military for about eight years," my training instructor told me at an amnesty hour visit. "I know that not all of the things we do here in training make sense. I am not preparing you to be a professional marcher or bed-maker or toilet cleaner. My mission is to prepare you should you be called on to serve our country in a war zone. You need discipline if you are going to survive a war scenario, trainee."

Just when I started to relax, feel a little better, and see my MTI as a friend, he'd let us know our time was up. "I don't know why you all are still sitting here. Don't you have a bed to make? I SAID GET OUT OF MY DAMN FACE, AND GO MAKE YOUR BED NOW!"

Okay, good talk, I'd say to myself leaving the dayroom.

During one of the amnesty hours, he told us all about how he earned his Purple Heart. That story stuck with me, and it went a long way in building my respect for him and deepening my sense of pride in the military.

He was also responsible for my single most embarrassing moment at Basic. During mail call, the MTI would either open our packages or make us open them in front of him to be sure we weren't receiving contraband. One time my mom sent me underwear. But not just any underwear. She sent me lacy panties and bras.

"What's in the package, Trainee Dumb-Ass?" he asked.

I hesitated.

"Trainee Clark, did I stutter?!" he screamed.

Slowly, I held up my panties.

"WHERE DOES YOUR MOTHER THINK YOU ARE? IN A VICTORIA'S SECRET TRAINING CAMP? THESE THINGS BARELY HAVE A CROTCH!" he screamed so loudly that I think my mom might have heard him hundreds of miles away.

After just a few short weeks in Basic, I transformed. All the things I struggled with early on—getting up at the butt-crack of dawn, enduring endless marches, doing push-ups until my arms felt like lead, and remaining stoic while a training instructor screamed in my face using my nostrils like a fleshy megaphone—became familiar, even normal.

When the MTI would get mad at us, he'd say, "I'm tired of looking at your stupid face. Why don't you get on your face and push Texas a little closer to hell?"

This phrase meant push-ups for punishment.

I grew to love this place with its structure, camaraderie, and fast pace. But Basic isn't a permanent home. Its purpose was to prepare me for transition into the military and my first assignment. And, at least in the short term, it readied me for graduation, which quickly approached.

Chapter 6

Finishing Strong

"The most rewarding things in life are often the ones that look like they cannot be done."

~ Arnold Palmer

WHEN I ENTERED Basic, I thought—wrongly—that I knew everything. "I know how to make a bed!" No, I didn't. "I know how to fold my clothes. I've been doing it forever." No, I didn't. The MTIs broke me down so they could rebuild me, emptied me out so they could fill me up, reminded me of my ignorance so they could teach me.

During Basic, instructors try out trainees in various positions of leadership. The most prestigious role is that of dorm chief. The dorm chief is responsible for the entire flight. Underneath the dorm chief are a handful of element leaders who are responsible for leading a row of people in their command. Essentially, the element leader makes sure that all the people in a particular row are squared away with everything they need.

One morning, the training instructor looked over our flight of trainees while we stood in formation.

"Okay, Trainee Dumb-Ass," he said, shaking his head. "I regret this already, but I guess I'd rather have you leading the element than Trainee Fat-Ass. You're up. Prove me wrong by getting this right."

"I'm getting promoted?" I asked, a smile spreading across my face in genuine excitement.

"Are we friends now?" he shouted back at me. "What the piss, Trainee! I want to drop-kick you in the back of your fucking throat!" he screamed, spit spewing from his beet-red face.

And that's when I knew I was his favorite.

I tend to give people the benefit of every doubt. Sure, his words may have been rough, but I thought I saw a glint in his eye indicating a suppressed smile. I'd broken the code. At home, my mom used to say stuff like that to me all the time. Well, not those words exactly. Whenever a child gets to feeling a little too big for his britches, Black moms across the country use phrases like, "You smellin' yourself?" and "Keep playing and see what happens," as a way to check the child back into reality.

During week six of Basic, we took part in Basic Expeditionary Airman Skills Training, known as BEAST Week, which was like a miniature version of a deployed combat scenario. We spent most of the week shooting, cleaning our weapons, and practicing in simulated combat environments. Training couldn't get much more real than this. It pushed us to the limits to see if we would break under stress. Some cracked. For me, ironically, this week stood out as the best time I'd had in my military career. I absolutely loved it.

At the beginning of Basic, I must have asked myself ten times a day, *What the hell am I doing here?* But by week three, I could eat, march, exercise, fold clothes, make a bed, learn in class, excel in tests, operate on little sleep, and suck up verbal abuse with the best of them. My surroundings evolved from familiar to enjoyable.

Not everyone, though, thrived in Basic. Four people from my flight washed out before graduation. One washed out the first week when she told the MTI that she hurt her ankle.

"Your recruiter sent you here when you weren't ready?" the MTI yelled at her. "Get out of my flight! You're going back. I can't believe they sent me a cripple!"

Another trainee suffered from depression and bipolar disorder. She lasted a couple of days before washing out. Another one washed out academically, unable to pass her tests. The final trainee got sent home because she kept casting spells on people.

One night, I woke up and found her sitting with her legs crossed staring at me.

"Wha…" I started. "What are you doing?"

"I can't sleep, so I've just been watching you," she said as if that would somehow make the situation less weird. Then she added, "Did you know you chew in your sleep?"

I did not miss her one bit.

Basic isn't merely physical conditioning. During this time, we continued to take classes, completed batteries of tests, and finalized our military job assignments. And that's when I realized that my recruiter lied when he assured me that I'd work in diagnostic imaging.

Most of the other trainees had pre-assigned jobs. Their jobs were listed in their military contracts stipulating a job placement guarantee. I had no such guarantee.

While I filled in diagnostic imaging on my application, S.Sgt. Peters filled in the word "general," meaning whatever job was available.

Four jobs were available: supply, medical supply, security forces, and maintenance. Once I learned that diagnostic imaging was out, the person assigning jobs asked me to rank my preferences from the four. After I took my first aptitude test back in Canton, I remember S.Sgt. Peters telling me that I could go in multiple directions. Except for one.

"Stay away from aircraft and wrenches," he told me. "You have no mechanical aptitude."

Good to know, I thought at the time.

I kept that in mind as I made my choices. I put security forces at the top of my list, considering a job in security had at least some connection to my first passion of criminology. Then I selected medical supply, since that at least got me closer to a medical education. Finally, I listed supply, the job I could not see myself doing.

So of course, they put me in supply, my last choice.

Graduation loomed right ahead of us, the moment where we trainees would become Airmen! Our training instructor gave us time to invite family to the ceremony, telling them unless we did something to really screw things up, we would be completing our training.

When I got accepted into the Air Force at the MEPS, my mom and Alexis had to work, and Brandy was away at college. Unlike Here Daddy, There Daddy (my biological father) wanted me to get some experience and then go to college, but he did not approve of me making a life-changing decision to join the military at such a young age. But on graduation day, both my mom and There Daddy came to show their love, support, and pride.

The day started with us doing a 5K run through the streets of the base with family members lining both sides to cheer us on. I've never run the Boston Marathon, but I can't imagine that being a bigger rush than our short run surrounded by our loved ones. While we ran, we sang cadences to keep rhythm, and our hearts swelled with pride.

Five weeks into Basic, my knee popped out on a march. It literally popped out. If I called attention to it, I would be recycled, meaning I'd be thrown out of my flight and have start all over again. *No fucking way*, I told myself. *I started this, and I'm going to finish it.* On that march, I popped my knee back into place and kept going. It swelled like a watermelon and hurt like hell, but I kept my eyes on the prize: graduation. Now, even weeks later, I felt my knee throbbing as I ran. I shut out the pain and kept going strong. I'd worked hard for this moment, and nothing would keep me from squeezing joy from it.

After the run, we trainees headed back to the dorm to shower and dress. Looking in the mirror before I got dressed, I remembered where I'd started. I'd entered Basic Military Training weighing ninety-four pounds. By the time I finished, my body had bulked up to 125 pounds of raw muscle. When I got there, I didn't know what abs were. But fed the combination of hearty meals and physical exercise, my body transformed into a machine, capable of doing things I'd never imagined. Not only did I look good, but ironically, the months of training were the happiest days of my life.

Graduation day meant we could finally wear our ceremonial blues, and we looked sharp. The weakest members of the training flight had washed out, leaving only the mentally and physically strongest of us to march out to the large stadium field with precision, our shoes shiny enough to blind onlookers.

Next, we marched on to the parade ground. As the commencement began, the commander explained to the hordes of family members what we trainees had gone through to earn this celebration. Watching the pride and love from the mass of loved ones brought tears to my eyes.

Standing as a neat, uniform flight of one, we earned our first award of the day: an airman coin. Then our MTI told us he was proud of us and he offered his congratulations. That act instantly erased any remaining hard feelings, embarrassment, and petty resentment I may have stored in my heart towards him. It communicated, "You are now one of us. You earned it."

From that moment on, the training instructor never again called us trainees. Never again would he yell, "What the piss, trainee?" if we fell short of perfection. I almost missed him calling me Trainee Dumb-Ass. Almost.

Then, like we had just weeks earlier, we raised our right hands and repeated the oath of enlistment. But this time, instead of wearing civilian clothes, we looked crisp and pristine in our dress blues. This time, when I raised my right hand, I did so in front of my mom and dad. Weeks before, I sounded like a kid trying to recite a poem I didn't understand and couldn't fully appreciate. This time, I spoke with confidence, authority, and pride.

After I said the oath in front of the multitude of witnesses, tears again streamed down my cheeks. That powerful, emotional moment will forever be etched as a highlight of my life. The ceremony evoked so many emotions that I finally let go and didn't try to wipe my face. And I knew somewhere in the sea of faces, my mom and dad scanned row after row of airmen, looking for me at that same moment.

I had done well in high school, especially once I started pursuing my passion. As great as I'd felt about making the grades in high school, this accomplishment felt monumental. I had done something on my very own. Enlisting was my first adult decision, and I did it at seventeen. I'd downed donuts and held my water to get there. In the process, I'd become capable of ignoring pain and pushing myself to new limits.

I had learned military history, endured endless screaming, run like the wind, and become part of a team.

Once the ceremony was over, we waited on the parade ground for our family members to find us. Which proved challenging. I felt like I was back at a high school dance. Everyone around me found a partner quickly except for me. I stood in the crowd, looking all over for my folks, but as the mob thinned out, I started to panic a bit. *The biggest day of my life, and I'm standing here alone*, I thought. I wiped away the tears of frustration and continued goose-necking around heads in the crowd in hopes of seeing my mom and dad.

And then, as weird as it sounds, I thought I saw my mom's ass.

"Michelle?" I shouted across the open space.

The woman turned, and then I got a better look. My eyes followed that ass up to her face. I'd never seen a more beautiful face in the world than when my eyes locked on my momma! Next to her stood my dad, his face beaming with pride. We rushed together in the middle of that field, holding on and crying like so many of my fellow trainees were doing with their own families.

We knew this little family reunion would be short-lived. This was Friday. We had the weekend together. On Monday, I would start tech school, where I expected the real fun would begin.

Chapter 7

The Two Faces of Tanga

"When dealing with two-faced people, it is difficult to know which face is uglier, the real one or the manufactured one."

~ Matshona Dhliwayo

MY TECH SCHOOL didn't require a long commute. I stayed in San Antonio and entered training just down the road.

I learned shortly before the graduation ceremony that I still hadn't earned the right to call myself an airman in the United States Air Force. That title would come only after completing more tests and on-the-job training. That's when a trainee becomes "operational" and earns the name airman. Until I passed tech school, I could still wash out with nothing to show for my efforts.

Tech school is the military equivalent to the freshman year at college. It was there that I met new friends, some that I hold dear to this day. Together we explored downtown San Antonio and did what most college freshmen do. I did well on my tests during my six and a half weeks there. Once I completed that, we had another big ceremony where our accomplishments were recognized. Finally, I could say, "Hi, I'm Airman Clark from the United States Air Force!"

When my first orders arrived, I learned I was to ship out to Holloman Air Force Base in New Mexico. When I told an airman who'd been around the block a few times that I'd be heading to New Mexico, he shook his head.

"You don't want to go there," he said flatly. "Holloman. It's in the name. That place is hollow. There's absolutely nothing there. You'll lose your mind."

"Well, what am I supposed to do?" I asked, assuming since the paperwork didn't say "requests" I had no options. "I have orders," I said, stressing the final word.

"Oh, that's easy," he responded. "Find someone to switch with you. The Air Force doesn't care if *you* go to Holloman. They just need a warm, trained body to report there. Ask around. Maybe you'll find someone who actually wants to go to New Mexico."

I asked around, trying to hawk my orders like I was holding hot tickets to a Beyoncé concert.

"Who wants fun in the sun and the gorgeous white sands of New Mexico?" I asked everyone I met.

To my amazement, a girl from my flight spoke up. "I do! I have family in Arizona, and that'll put me closer to them!"

"I don't know," I said hesitantly, fearing she might have orders for Thule Air Base 700 miles north of the Arctic Circle. "Where're they sending you?"

"Langley," she answered, meaning Langley Air Force Base (AFB) in Virginia, a couple of hours from Washington, DC. Nearly every member of my flight from tech school had orders to go to Langley. While I can't say that Virginia came anywhere near the top of my list for my dream station, if I went there, I'd at least know some people.

Before I could say anything, she made the decision easy for me.

"I'll give you $2,000 to change orders with me! Please?"

"Sold!" I said.

I spent that money on airfare home to see my family. Then I bought Christmas gifts with the money left over.

My orders were to arrive at Langley AFB on January 11, 2011, and since the Christmas holiday approached, we had exodus where we could take some leave to visit family. I also tapped into the Recruiters Assistant Program, which gave me some extra time off. In total, I had four and a half weeks of leave before I had to arrive at my first duty station.

Under the terms of my leave, I had to check in with my recruiter when I got back home in Canton. But I would have stopped by to see him even if I weren't required.

"You lied your ass off," I said as I walked through the door of the recruiting station. "Almost everything you told me was a lie. I'm not in diagnostic imaging. I'm in supply," I began before launching into all of the other lies and half-truths he'd told me to get me to sign.

"Okay, look," he said when I finished blowing off steam. "You're supposed to check in and work a program with me throughout your leave. How about you just have a good leave, and I'll say you completed the entire program, okay?"

"Yeah, well, that's the least you can do," I answered.

True to his word, S.Sgt. Peters didn't bother me during my time at home, and he put in a good word for me to boot. That left me with about thirty-five days with my family and friends that had returned home from college for winter break.

But all good things must come to an end. After a fantastic break at home, I flew out to Langley AFB to start my full-time job in supply on January 11. Supply is like logistics. I told people that I was like the UPS of the Air Force. I tracked every piece of Air Force equipment of high dollar value—things like computers, phones, crypto items, classifieds. In addition to tracking equipment at Langley, I managed five to eight base-level equipment accounts.

Unfortunately, tech school did not prepare me for my role. My tech school instructor covered basic material, but she didn't take the time to teach the procedures used at Langley or Scott.

"There's a zero percent chance that you're going to be sent to Scott or Langley," she said dismissively. "They don't send new airmen to headquarters. I'm not going waste time teaching you how to manage a major hub when you're heading to some small, regional base in the sticks."

Ironically, most of my class got orders for headquarters, and we would find ourselves lost from day one. When I first told my instructor that I was heading to Langley, she lowered her head.

"I'm so sorry," she said sincerely. "Fortunately, you're smart. You'll be able to pick things up."

Either by default or design, my first supervisor at Langley, S.Sgt. Tanga, handed me her continuity book, which was like her own personal cheat sheet for various procedures. She told me to scan every page to create my own book. In the process of replicating her book, I read each sheet. She really knew her stuff, since she'd been in her role long enough to have learned tons of shortcuts and helpful tips. By the time I finished making a copy of the book for myself, I had the manual of one of the most knowledgeable supply supervisors I'd ever known.

Little by little, I picked up what I needed to know, mostly through on-the-job training and tips from more experienced airmen. Then I took the First Term Airman Course (FTAC), a two-week course required of all new airmen to reiterate what we'd learned in Basic in addition to in-processing the base, learning the different programs, agencies, and military benefits. Next, I was assigned a set of career development course materials, which included a series of books and tests to accelerate learning about my job. Those resources gave me confidence to transition from trainee to a full-fledged, operational airman.

Just as corporate employees have titles designating their relative importance to the organization ranging from hourly workers to CEO, the military follows ranks and chains of command that many civilians don't understand. In Basic, I was a trainee. In other words, *I was less than nothing*. Trainee means "maybe you'll make it, but only time will tell." Once you complete training and testing, you move up to E1, which is when you earn the title of E1: airman basic. If a trainee is less than nothing, an airman basic is nothing. From there, the ranks move up the ladder as follows:

E2: Airman (where you earn a chevron on your uniform)

E3: Airman First Class

E4: Senior Airman

E5: Staff Sergeant (S.Sgt.)

E6: Technical Sergeant (T.Sgt.)

E7: Master Sergeant (M.Sgt.)

E8: Senior Master Sergeant (S.M.Sgt.)

E9: Chief Master Sergeant (C.M.Sgt.)

Since I signed up for a six-year stint, I went from airman basic directly to airman first class.

Promotions into the ranks of S.Sgt. to C.M.Sgt. are comparable to corporate positions of supervisor, assistant manager, manager, store manager, and district manager.

Officer ranks begin at second lieutenant and continue through general. In most organizations, you could consider these leaders as vice presidents through COOs. A squadron commander is typically filled by either a major or a lieutenant colonel, which is something like a senior vice president.

I share that to show you my rank and position relative to the squadron commander. At that time, my commander at Langley AFB was Major Skyler, a man I thoroughly respected and appreciated. He loved the Air Force, but he also cared for his airmen enough to get to know us and shepherd our careers. I can't say enough good things about him as a commander, leader, and human being. Sadly, I found leaders like him a rarity. My admiration for him grows to this day—that a leader at his level would care personally for someone like me at the bottom of the ranks.

My first supervisor at Langley AFB, S.Sgt. Tanga—the one who gave me her book to copy—knew her job well. But she was no leader. Instead of motivating and inspiring us, she tried to be cool and make friends with her direct subordinates. While she was much older than the rest of us (meaning she may have been thirty while the rest of us were under twenty), she partied harder than those of us who reported to her.

"You and Chris should come with me to Harold's house this weekend," S.Sgt. Tanga offered. "It's kind of an office thing that most of us do on the weekends. It'll be fun, and you'll meet some people. Besides, I'm sure you could use a few drinks to let your hair down after all you've had to learn this first month."

"Sure," I responded. "That sounds like fun."

Good, I thought. *Chris and I have never been to a party together before.*

I met Chris in tech school, and the two of us started dating. Chris acted like a typical jock with more than a dash of frat boy in him. He was a people magnet due to his outgoing, carefree attitude. But those same traits attracted critics, because Chris had a hard time adjusting to the military lifestyle. To no fault of his own, my supervisor enabled him a lot. It didn't take long for Chris to realize that the Air Force he joined was different than what he'd hoped for. He entered the Air Force full of dreams and goals, but he soon grew disillusioned—which led him to binge drinking and regular partying. Chris was the life of the party, until he tipped over into drunk and became annoying. His last name was Baudette, pronounced like the French word *baudet*, meaning jackass. When drunk Chris made an appearance, everyone called him *baudet*.

No one has ever considered me shy or timid, but being new in the military, I held back in social settings. I wanted to have fun, but I didn't want to get into trouble. I felt enough excitement at parties just watching Chris act a complete fool. Besides, he served as a great deterrent from wanting to drink.

The running joke in the military is that everyone meets the love of their life in tech school. They call it "tech school love." Like puppy love, it's not meant to last. From the first time I met him, I pictured the two of us beating the odds and becoming forever loves. Thank God we never married. Chris was the relationship I needed to discover everything I did *not* want in a partner. While we had amazing ups, we had even more spectacular downs. In hindsight, I fell in love with the concept of love.

Many things changed over the course of our relationship, except for one thing: Chris partied all weekend long.

Chris and I showed up at Harold's party around the same time that S.Sgt. Tanga arrived with a few bottles of liquor. For some reason, I didn't drink at all that night. Had I been drinking, I wouldn't have noticed my supervisor flirting with my boyfriend. But sober, I couldn't *not* see it. Watching my own supervisor flirt with my boyfriend pissed me off so badly that I left the party.

Several months later, I learned that Chris tried to have a threesome with S.Sgt. Tanga and one of my friends that night. Had I known that at the time, I would have had an easy decision to make. But what I did hear about the party still disturbed me. Chris and my supervisor spent an hour lying in bed together—with Chris, supposedly, passing out from alcohol consumption. Even if that were the "only" thing that happened that night, I was still pissed.

"What the hell?" I said to S.Sgt. Tanga back in the office on Monday.

"Look, it is what it is," she said with no trace of remorse or even embarrassment. "And whatever happens at Harold's stays there. We never bring it back to work or talk about it. So if you have a problem, we'll talk about it later, Kia. And if you have a problem, just know I'm your supervisor, and I write your enlisted performance report. Just as fast as you got into the Air Force, I can get you out."

That was the first threat I received from a military boss. It wouldn't be the last.

I kept my mouth shut. Obviously, that did not make for a positive, harmonious working relationship. But she had rank on me and served as my supervisor. I'd been trained well. I would follow orders. That didn't mean I had to like or respect her. And I didn't.

Ironically, S.Sgt. Tanga had a fiancé in Texas.

While at another party with Chris a few weekends later, I wanted to go back to the dorms, but Chris was far too drunk to drive. I found a friend, Tony, who was leaving, so I asked for a ride back with him. As we were leaving the party, we saw S. Sgt. Tanga pinned up on the side of the apartment complex with a leg wrapped around the guy while they swallowed each other's tongues. Since the two of them were right in front of our car, we had a front-row seat. And from that seat, it looked like the guy's pants were undone and loose around the waist. I said nothing, but as soon as we got in the car, Tony said, "Typical Tanga." Nothing more was said about that.

The next day, I got a phone call from someone who wasn't at the party asking me if I saw anything happen between Tanga and another guy.

"No," I said repeatedly. "I'm sorry. I didn't see anything." Not *my circus, not my monkeys*, I thought to myself.

But then the caller broke down into tears.

"Kia," he said, sounding broken, "my best friend is about to marry S.Sgt. Tanga. If you know anything about her messing around, please tell me. My friend is about to move up there and start a life with her."

"I, um," I started. "I didn't stay long at the party. I left early. So I don't know. But look, maybe you should tell your friend that if he's got a gut feeling about her, maybe he should just follow his gut."

When my supervisor learned that I had spoken to her man, she gave me hell.

"Oh my God, Kia!" S.Sgt. Tanga called me. "Why would you say anything? You're about to ruin my life. I cannot believe you did this to me!"

I explained that I didn't speak to her fiancé, nor did I say anything to anyone. All I had done was talk to a mutual friend who called me to say her fiancé was getting cold feet.

"I told him that I didn't know anything about marriage, so maybe he should just follow his gut," I told her.

I was still eighteen years old, and I knew that I didn't have all of life's answers. A friend of S.Sgt. Tanga tried to clue me in.

"You broke the *girl code*," she told me.

"I didn't know we had a girl code," I said honestly.

"Look," she said, explaining the facts of life to me. "We all know that S.Sgt. Tanga sleeps around. But that's her business. It's okay that we all know her business, but we don't talk about it, especially not with her fiancé."

Good to know, I thought to myself. *There's a girl code, and I guess I broke it by not sticking with a lie. New rule: follow the code designed to protect liars and cheaters. Got it. But anyway, I didn't even talk to her fiancé!*

I was barely two months into the military. I didn't know these people well, and I'd never imagined that because of the guy I dated I'd be thrown into a life of partying, drinking, drama, and deceit.

Long after the fact, I learned that S.Sgt. Tanga and my boyfriend shared a bed while passed out together. Much later, I learned that they also slept together that same night. Had I known it at that time, would I still have been expected to keep to the code? Was there a code that said, *Don't sleep with the boyfriend of one of your employees?*

Even though I breached the girl code, S.Sgt. Tanga patched things up with her fiancé and got married just the same. Not surprisingly, it wasn't long before they were at the courthouse going through a divorce. Care to guess why? Yeah, she kept sleeping around.

This experience taught me an invaluable lesson, though. From day one, Air Force training drilled this mantra into our heads: "If you see something, say something." While that sounded great in principle, the military structure made it impossible to achieve. How the hell could I "say something" when my supervisor slept with airmen in her charge, provided alcohol to minors, and threatened my military career if I spoke up?

If you see something, keep your mouth shut became one of my own rules—one that I planned to keep.

Chapter 8

Man Out!

"Is this what growing into an adult woman is—having to predict and accordingly arrange for the avoidance of sexual harassment?"

~ Candice Carty-Williams

BACK IN SUPPLY, much of my professional development was shaped by Chief Master Sergeant of the Air Force (CMSAF) Cody. The CMSAF serves as the highest enlisted level of leadership in the Air Force. The person in that role advises on all issues related to the welfare, readiness, morale, and proper utilization and progress of the enlisted force. Even though Chief Cody worked at the Pentagon, his influence reached all levels of the enlisted forces. He was big on professional development and making sure those under his charge understood the business from the inside out. He demanded professional military attire, which meant each Monday we dressed in our service dress uniform, the fanciest uniforms we were issued. We called it *blues Monday*. Looking back, this may have been Chief Cody's way of bringing us out of our weekend mindsets and back into the business at hand.

I didn't mind wearing my blues, even though all the women looked like flight attendants and the males resembled captains and co-pilots. Instead of working a mid-air flight though, we sat in what could have been any corporate office with cubicles and computers.

One morning while working at my desk, I mindlessly slipped my foot out of the back of my shoe and bobbed it up and down by flexing my toes.

"Keep doing that with your shoe," said S.Sgt. Jacobs from the cubicle across the aisle.

Not sure his words were directed at me, I turned around and said, "I'm sorry?" to see if he was talking to me.

"Your shoe," he repeated. "What you're doing with your shoe. It's turning me on. If you keep that up, I'm going to come over there, bend you over your desk, and fuck you from behind," he said very nonchalantly.

"Excuse me?" I said, completely stunned. As a hummingbird full of nervous energy, I'm constantly moving. While I hadn't been consciously aware I'd been doing anything with my shoe or foot, I now made keeping my feet still my number one job.

But when I turned back to my work, I felt my face flush, first with embarrassment and then with anger. *What the piss?!* I wondered to myself. *Who the hell says something like that out loud, especially out loud in an office full of people?*

After fuming for a while, I went to talk to T.Sgt. Gillis.

"So then he says," I relayed word for word, "'I'm going to come over there, bend you over your desk, and do you from behind.'"

"He said that?" T.Sgt. Gillis asked.

"Well, no, sir," I answered. "He used the 'F' word."

"Okay," he replied flatly. "Did you stop playing with your shoe?"

"What?" I answered. "Well, yeah. I stopped. But did you hear the rest of what I said?"

"Yeah, I heard it. Listen, Clark," he started, "if you're going to be offended by literally everything people say to you, then you probably just need to get out. Cuz I'll tell you right now, it only gets worse if you get deployed. It's a male-dominated Air Force, so you better get used to it," he said with a finality that conveyed that since I now knew the true reality of Air Force life beyond the recruitment posters and platitudes of "see something, say something," I should just lighten up.

"Okay," I responded. "Good to know." On the outside I kept calm—my demeanor almost expressing gratitude for his sharing this wealth of new information. But on the inside, I felt humiliated, harassed, and broken.

I quickly learned that T.Sgt. Gillis spoke the truth about what to expect in the Air Force. Other airmen made similar comments to me, and I told myself that *This is just the way things are.*

You better get used to it. I could hear those words by T.Sgt. Gillis in my head as well. *Besides*, I told myself, *who's going to give me a different answer? I'm just a snack inside the boys' club.*

A few days later at physical training (PT), one non-commissioned officer (NCO) looked at me and said to my boyfriend Chris, "Damn, dude! You get to hit that every night?" Then he stared at my body as if I were a stripper on a pole, as if I had chosen to show up at PT, as if I should be thankful that he looked at me like he wanted to stick dollar bills in my sweatpants. I heard another NCO say, "I would so hit that," as others offered him a high-five.

I forced myself to take whatever shit they dished out. Only months earlier, I had joined the Air Force and took the oath of enlistment. I longed to make a difference, serve with distinction, and protect the freedom of this country. I wasn't going to let a few assholes make me question that decision. I determined that I would not be labeled the girl who "couldn't take it." I sucked it up.

Shortly after this incident, another sergeant in the office made himself known to me. S.Sgt. Morin seemed very plugged into us newer airmen, frequently checking in with us to see how we were adjusting to military life. On top of that, he came across as a stand-up human being and godly man who regularly invited us all to join him at church. Each Friday, he held mid-morning meetings in the breakroom to see if any of us needed anything.

"How are you all doing this fine day?" he asked cheerfully as he entered the office one morning. "I just want to see if you're doing alright, find out how you're adjusting, and learn if there's anything I can do for you."

It felt good to have someone so upbeat and positive in the office. And his invitation to join him at church seemed so much more wholesome than someone asking me to the bar.

"Just so you know," S.Sgt. Morin told me and other newcomers, "I like to meet with you together once or twice a week so we can get to know each other better and deal with things that come up before they become problems, you know?"

That sounded great to me, so three other female colleagues and I joined him at one of his voluntary newcomer meetings. My boyfriend, Chris, was new, too, but I didn't notice that he hadn't been invited to join us. Over time, one woman stopped coming. Then as more time passed, the other woman, Tracy, who'd been a friend of mine since tech school, dropped out, leaving just me and S.Sgt. Morin. Which seemed weird. Three people can be a group. But two people? No, that just felt strange to me.

"Where were you this morning, Tracy?" I asked her.

"Oh, I'm not going to those meetings anymore," Tracy answered without expounding.

"Why?" I pried.

"I don't have to have a reason," she shrugged. "I just don't mess with him no more."

I knew her well enough to understand she could say more. She lived her life as far away from gossip and drama as you could get. Her brief words didn't mean she had no opinion; it just meant that she preferred to keep it to herself.

But I kept pressing.

"But why?" I nudged.

"I just don't like him, okay?" she said, closing the door to any further discussion.

I kept going to S.Sgt. Morin's meetings. Learning customs and courtesies at Basic Training and having it reiterated at FTAC taught me that I should show respect to everyone, especially those of higher rank. I kept attending his group, even though my colleagues decided it was okay to disrespect him by blowing him off.

It didn't take me long to see the paradox of S.Sgt. Morin. One side of him was this church-going family man who offered his time to ensure his airmen learned the ropes. But the other side, the one I couldn't wrap my head around, was his creepy side.

During PT one morning, S.Sgt. Morin said in a loud voice, "Hey, I have an idea! Why don't you," he said, pointing at me, "and Hua put on French maid outfits and roll around in the mud for us? What do the rest of you think?" he asked the men around him.

"I think that's the best idea you've ever come up with!" someone yelled in support. "That would be so great! Sexy!" others said.

"Nah," I said as I waved my hand in the universal goodbye gesture. "I'm good."

Later, I asked Hua, "What's the deal with Morin? Some days he's all 'Praise Jesus,' and other days he's like, 'Why don't you mud wrestle for me'?"

"I don't really talk to him anymore," Hua spat out.

"Yeah," I agreed. "I've never seen that side of him before."

"Really?" she said. "Did I ever tell you that he used to try to show me porn?"

"Wait! What?" I said as I tried to wrap my head around the image of this positive, godly man showing a young woman porn.

"Yeah," Hua continued. "But that's not the creepy part."

Now I'm thinking: *You don't think THAT'S creepy? What the hell are you about to tell me next?*

"Then he'd say, 'Whenever I see an Asian girl in one of my videos, I picture your face on the woman's body.' How creepy is that?" she asked.

"Are you serious?" I asked. "That is *not* okay. Did you say something? Did you report him?"

"I told S.Sgt. Tanga that I wanted to report him," Hua answered. "She talked to him, I guess. She told him that if he did it again, I'd turn him in."

"Did he stop?" I had to know.

"Yeah," Hua confirmed. "This was like a year ago. But what he said about the French maid's outfit was just weird."

"I can't believe it," I said in shock.

"Watch yourself," Hua told me. "The only reason he left me alone is that he found someone else."

I stopped attending S.Sgt. Morin's group. It didn't take me long to figure out that I was his *someone else*.

Around that time, a coworker celebrated a birthday, and a civilian colleague baked him a cake. The cake was shaped like two breasts covered in white frosting with a huge penis ejaculating between them. Several of us took pictures of the cake, me included. While I found it disgusting, I sent it to a few friends back home saying I felt like I never left high school.

A few days later, our section chief ordered us to line up.

"Take out your cell phones," he ordered while we stood at attention in front of him.

Another sergeant walked through the aisles, stopping at each of us.

"Pull up your photo gallery," she ordered.

Once she found what they were looking for, she'd say, "Delete that photo. Now delete these two photos."

She made us delete all photos of the cake. Then our section chief issued one final order.

"We will not talk about this. Don't mention it ever. It never happened," he said.

With only around five months into the military, I figured: *Hell, you don't want me to talk about it? Talk about what? What are you talking about?*

But it wasn't over just yet. A couple weeks later, investigators came into our unit asking questions about the cake. But I didn't know they were investigators. They looked like civilians. When they asked about the cake, I thought this was a test of loyalty.

Playing in the back of my mind was the last "test" I went through, the one where I reported that a coworker threatened to bend me over a desk because I had the nerve to wiggle the shoe on my foot. I figured I'd failed that test.

"You better get used to it," echoed in my head. "If you're going to be offended by literally everything..." I remembered T.Sgt. Gillis telling me.

"I don't know what you're talking about," I told the questioners. "I never saw an inappropriate cake."

Not only did I learn to shut my mouth, but I also learned to get over things quickly, like S.Sgt. Morin's suggestion that Hua and I put on French maid outfits and wrestle. I'm naturally forgiving and tend to let go of things anyway.

I continued to receive sexually demeaning comments on such a regular basis, that I redefined what I considered acceptable. *Talk dirty in front of me? Fine. I can live with that. Talk dirty to me and about me? I don't like it, but I guess this is the Air Force way. I'll live with it, as long as you don't touch me.* My boundaries were pushed continually, and I kept letting them get pushed further. I could speak up and get threatened, made fun of, and told, "You're too sensitive!" Or I could ignore it, telling myself to suck it up since no one physically hurt me.

It wouldn't be long before I had to rethink that strategy.

One morning, S.Sgt. Morin sent me a text while I worked. At first, I had no clue who was texting me.

"You don't know who this is?" he answered in response.

"No," I repeated. "Who is this?"

"S.Sgt. Morin," the reply came.

Weird, since he sat not far away from me.

"How'd you get my number?" I asked, a little confused.

"Recall roster," he responded.

That made sense. The recall roster listed the names, phone numbers, and addresses of everyone in the office so you could contact them in the event of an emergency like a fire, tornado, or, more likely near the coast, a hurricane. We all had access to the recall roster, and we were required to keep that list on us at all times.

I didn't know what emergency made him reach out, and before I could ask, he texted again.

"I'm so scared. I'm nervous," his message read.

"What's going on?" I asked, wondering what happened to him.

"My man's out, and I don't know what to do," he answered.

I knew that he and his wife were expecting their second child any day now, a boy. He had told us with excitement about his good news several weeks earlier.

"Congratulations!" I texted back.

"No, my baby's not here yet. My man's out," he said again.

"I don't know anything about kids, but T.Sgt. Colon has kids. Maybe you can get in touch with him," I offered.

"No, my man's out," he repeated.

I'm racking my brain trying to figure out what he was talking about. Then it dawned on me. He's black. His homeboy just got out of jail.

"Your homeboy got out?" I texted.

"No, my man's out," he repeats.

"Are we playing hangman here?" I asked, totally lost about what he was talking about.

Finally, I shared the string of messages with a friend in the office.

"Any idea what he's talking about?" I asked her as she read through the chain.

"Girl, I don't know what he's talking about. Did you actually ask him?" she suggested.

I decided to put an end to this guessing game and sent him another message.

"What are you talking about?" I typed.

"Nothing," he replied quickly. "I need to go to the bathroom. Is it open?" he asked.

I checked the bathroom, a one-seater with a lock from the inside. It was locked.

"Let me know when it's open because I really got to go," he texted before I could reply. "I'm like shaking. I got to go."

You don't eat the way I do without occasionally getting the shits like his message suggested. It can get so bad you start sweating. I felt his pain.

"I know how it is," I answered.

"I'm scared," he texted. "What if the commander comes by? I'm shaking here!"

Thinking he's ready to shit his pants, I head over to his desk.

When I got there, S.Sgt. Morin stood up and his pants were unbuttoned and opened. His penis was in his hand, and then he started coming at me.

Cubicles were squared-off into a quad holding four desks with a small opening for entering and exiting. S.Sgt. Morin wasn't alone. Inside the cubicle was another airman who had his back turned away.

Without saying a word, I turned and headed straight to my friend's desk.

"I got to talk to you," I began. When I finished, I was still in shock, and I asked her, "What should I do?"

"What you're going to do is tell somebody," she answered immediately. "Like now, right now. You need to tell your supervisor."

I was eighteen years old, hardly old enough to buy a pack of cigarettes, and still years away from being able to enter a casino. Yet apparently, I was old enough to be told at work that I could be bent over a desk for shaking my foot, and now I could have a man expose himself to me.

Against my better judgment, I decided to break my silence.

Chapter 9

At Least It Can't Get Worse, Right?

"You never know when your life is about to change. You never know when one decision will dramatically impact your life and change the course of your destiny."

~ Dani Johnson

I RUSHED TO the office of my new supervisor, S.Sgt. Golden, asking if I could report something to him. I don't know if my choice of words or the look on my face conveyed urgency, but he grabbed his supervisor. My friend stayed by my side, and the four of us went into the break room to talk.

Shaking with anger and shock, I detailed what had just happened. If *WTF* had a voice, it sounded like mine. If it had a face, it mirrored the look I'm sure I wore by the time I finished talking.

"I had no idea what was going on," I finished, still trying to calm myself down. "One minute I'm working, and the next, he's whipping out his penis and trying to grab me."

"Let me see your phone," my supervisor said, holding out his hand.

Thank God, I thought. *They'll read it for themselves and put a stop to this behavior.*

As we all stood in the break room, I watched as the two sergeants scrolled to the first message and began reading. In seconds, they started laughing as they continued to read. A few times, they laughed so hard they had a hard time breathing. When they finished, my supervisor put my phone down on the table.

"Just how stupid are you?" he asked, his own boss laughing so hard he couldn't look at me. "I mean, I knew what he was talking about way up here at the top of the message when he says, 'my man is out.'"

"Um," I stuttered. "Okay, I'm not a guy. That meant nothing to me."

"Alright," he said. "Go to lunch. We're going to go talk to the higher-ups and see what they think. We'll talk about this later."

To say I felt confused would be an understatement. I'd known a lot of men, and I'd known even more immature boys. Never in my life had any of them said "my man is out" and pointed an erect penis at me while trying to grab me.

Was this acceptable in the Air Force? Was this just another thing I was expected to 'get used to'?"

Normally food gives me a reset. Not this time. My stomach hurt, and I felt like I needed to puke. Making it worse, while I sat at lunch, S.Sgt. Morin began texting me again.

"Where you at?" he texted. "I'm in the bathroom. I'm scared."

I ignored him, furious that he wouldn't leave me alone even now.

Before I could swallow more than a few bites of food, S.Sgt. Golden called me.

"Get back here ASAP," he said. "Come find me when you get here."

I had been sitting at lunch for six whole minutes and hadn't been out of the office for fifteen. But it felt like hours as my stomach became a giant knot. I couldn't wait to get this behind me.

"Attention," my section chief, M.Sgt. Sheraton ordered as soon as I entered her office. Waiting inside were S.Sgt. Golden and his supervisor T.Sgt. Colon, the two sergeants who had read the text exchange. In the Air Force, there are three times you're put at attention. First, when you win an award. Second, when they're reading a decoration. Third, when you're in trouble. My mind swirled as I tried to figure out what award I could be earning.

Immediately, my section chief began reading a letter to me. And just as quickly, I recognized from her demeanor and words that this was no award. It was an official reprimand.

"You are hereby counseled for texting on your phone inappropriately at work," she read aloud.

The Air Force has no shortage of protocols around disciplinary action. Typically, if an airman commits a minor offense for the first time, she'll receive verbal counseling. As the name suggests, an airman is told by a supervisor what she did wrong, followed by some version of "don't let it happen again." The next step in progressive discipline is a "letter of counseling" if the problem persists. Higher up on the ladder of severity is a "letter of admonishment" for more serious offenses. I'd never seen these being used. The most severe level of discipline is a "letter of reprimand."

All these types of discipline are called "paperwork" for short. Paperwork is bad, because it goes in your file (known as your personnel information file) and stays with you as long as you remain at that base. It's like having a criminal record.

For my part in receiving text messages from S.Sgt. Morin, my leadership viewed my behavior as so grievous that they gave me a letter of counseling. Not verbal counseling. They skipped over that to lower the boom on me. S.Sgt. Morin, who exposed himself to me and then charged me, received a letter of reprimand.

I'm eleven months into the military, and I've committed my first offense for inappropriate texting at work.

"I'm sorry, do we have a no cell phone use at work policy?" I asked, trying to understand the nature of my reprimand.

Instead of answering, she made a face.

"There is no policy that I'm aware of, and I've read all of the policies," I continued. "And I wasn't on my cell phone when I should have been working. Sergeant Morin texted me that he needed help. I did nothing wrong, and I certainly didn't text anything inappropriate," I spoke, by this time tears falling down my cheeks.

She offered no explanation except to scream at me. I couldn't understand what she was saying, except I knew she wasn't going to change her mind.

What the hell is wrong with these people?! Some asshole shoves his dick at me, and I get punished? I'm the victim here, but I get punished? I cursed the empty sidewalks as I walked home later that day. *Why did I open my mouth? I can't win for losing!*

Langley is a large base, but people talk. Some airmen in the office witnessed what happened, so the story bounced around quickly, including that I'd gotten written up. It didn't take long for Squadron Commander Major Skyler to catch wind of the story. He had zero tolerance for the kind of behavior that S.Sgt. Morin displayed.

The squadron commander called me in to inform me that S.Sgt. Morin would face a court-martial.

"You will be called to testify at the upcoming court-martial," he informed me after he'd called me to his office.

To be honest, I didn't even remember what court-martial meant except that it was bad. I mean, I'd learned about disciplinary action in one of my training classes, but why would I retain knowledge about something I never planned to need to know? It's like making a list of all the things I'm going to buy when I have a billion dollars. *Court-martial*, I thought, *on what planet is that really going to happen?*

Once he explained what I needed to do, I agreed.

Good, I thought. *That asshole should be held accountable.* I tried to imagine any other occupation where this sort of thing could happen at work. In every scenario I could imagine, if you got your penis out at work, it's not the other person that gets fired.

Finally, I'm going to see some military justice, I thought.

Instead of justice, I experienced what the military uses to ensure the conformity and loyalty of its members: shaming and hazing.

"What the hell is wrong with you?" I heard a version of this sentiment each day. "Why are you doing this to a good Christian brother? How could you be so terrible to your own people?"

Okay, back up, I'd think to myself. *Do you remember in the Bible where Jesus whipped out his penis while delivering the Sermon on the Mount? No? Me, neither! And as far as "my own people," all the people I call MY PEOPLE keep their pants pulled up at work!*

Some attacked my morality in addition to the religious and race cards.

"You're telling me you've never seen a penis before?" too many people asked. How often did I hear this? Often enough that I acquired my first nickname since Basic: Penis Girl.

"Don't talk to Penis Girl," I heard over the next couple of months. "She'll end up getting you in trouble."

Each time someone said these things to me, I'd come back with something like, "But I never asked to see him expose himself. You get that, right? He forced that on me."

One thing I'd learned from crime shows on TV is that when a woman is a victim of a sexual crime, the defense attorney blames the victim. "You provoked him" or "What were you wearing?" or "What did you do to stop it?" And it felt like those working around me were playing S.Sgt. Morin's defense attorneys, treating me like garbage because I opposed working in such a highly sexualized work environment.

I went back to keeping my mouth shut and expecting little from those around me. Even when I asked my supervisor a work-related question, he would walk away from me as if I did not exist. More than once, I returned to my desk after a break to find my office supplies and even lunch had been thrown in the trash. This wasn't just a random event or the acts of a few; the entire culture seemed supportive of treating me like a leper.

While I was treated like the bad guy, some treated S.Sgt. Morin like the victim. Coworkers were assigned to S.Sgt. Morin detail, meaning they drove him to appointments and brought him food. Flight Chief Mr. Coates eventually held a meeting where the airmen taking care of S.Sgt. Morin received special recognition via letters of appreciation.

The squadron commander told me I was doing the right thing, but I had earned a letter of counseling.

Recently, I reconnected with a former colleague in the Air Force named George. He told me that the office environment became so toxic that he contemplated suicide. Instead of killing himself, he slipped into a bottle.

The day before I testified, two other airmen came to me and said S.Sgt. Morin had sexually harassed them.

"Why didn't you say something?" I demanded.

"You're kidding, right?" one of them answered. "Look at how everybody treats you. There's no justice in the boys' club."

Before the hearing, I cried uncle to the squadron commander.

"I can't do this," I cried. "Please call off the court-martial. I don't want to press charges. It's fine. Let's just drop it, please? I didn't ask for this," I begged.

"I'm sorry, Airman Clark," he said with genuine compassion. "In my position and based on my core values and the values of the Air Force, I can't let this go. He needs to go through this court-martial. And that means charges must be brought on him. We need to hold him accountable. What he did is not acceptable under any circumstances."

Instead of responding, I tried to hide my tears, not wanting to show my brokenness.

"Hey," he said kindly. "Are others giving you a hard time? If anyone gives you flack, you need to let me know. Because I will fall on them like a brick house."

I wanted to say, *I'm being treated like I'm the pervert! What they are doing to me is worse than what Morin did to me. He was just one person, but now I have everyone against me! Do you know what it's like to be called Penis Girl every day? Do you know how often I've had to fish my belongings out of the garbage? Do you have a clue about what it feels like to be called an Uncle Tom by people I thought were my friends?*

Instead, I said nothing. I couldn't tell him. If complaining about sexual harassment caused everyone to gang up on me, I couldn't imagine what they might do if I ratted them out. They already hated my guts, and I decided anything else I might say would only make it worse.

The hazing continued for two years.

It wasn't until 2013 that the Air Force instituted a special victims' counsel for victims of sexual assault and rape. If you were a victim, they assigned you a lawyer, so you didn't have to go at it alone. But at the time of the S.Sgt. Morin court-martial, it was just me against my entire unit.

My self-esteem continued to plummet. Chris and I had a messy breakup. Initially, I'd dated him with doe-eyes and naivety. He could be the life of the party, but as the saying goes, he never wanted to leave the party. He and S.Sgt. Tanga continued flirting with one another, and I kept forgiving him. He slept with S.Sgt. Tanga, I learned later.

But even before I learned that news, I saw him trying to sleep with anything that moved. As my self-esteem dropped further, I started believing that I couldn't do better or shouldn't expect better than what he offered me. Even when he got black-out drunk and went on rampages, I stayed. He cared nothing for me and treated me like dirt. But I stayed with him rather than being completely, utterly alone.

Then he broke it off with me. Chris—who left me stranded in a broken-down car, driving past me because he planned to go fishing and couldn't be bothered, who once wrestled me to the floor, sat on top of me, grabbed both my hands, and slapped them together until they were red and numb, whose parting words weren't to me but to his friend saying, "I'm done with that. Hit all day long if you want"—broke up with ME.

Just when I didn't think I could sink lower or feel any more alone, Chris broke up with me.

I doubled down on putting on a happy face and staying upbeat. When I'd pout as a child, I remembered hearing the phrase, "Nothing like a frown to make a pretty face ugly." I tried to pick myself up, willing myself to rise above. I stayed quiet, keeping my head above the waves while waiting for the storm to pass.

When Chris started dating one of my friends, I never said a bad word about him. In fact, I defended him. When he got drunk, arrested, and charged with underage drinking, I feigned surprise. When he was at the center of an alcohol-related incident on the base and my friend made a report against him, I still kept quiet.

"Chris?" I said in response to questions about his character. "Who doesn't like Chris? He's funny. I can't really think of anyone who doesn't like him."

If Chris were to get kicked out of the military, it would be based on his behavior and not anything I would throw on him. And what I said was true: almost everyone loved Chris, just like most people loved S.Sgt. Morin.

Thinking back at how mad I'd been when Morin's victims left me standing out there all alone after he'd done similar things to them, I feel ashamed now. I should have told the investigators what I knew. But at the time, I told myself, *You spoke up twice now, and look how that worked for you?*

Fortunately, Chris got thrown out of the military without my help—I didn't even need to make a case against him. And unfortunately for me, people assumed I had a hand in it.

Chapter 10

Firsts

"When you fall head over heels for someone, you're not falling in love with who they are as a person; you're falling in love with your idea of love."

~ Elisabeth Rohm

S HORTLY AFTER TURNING nineteen, I wanted nothing more than to be back at home, lying in my old bed, and reminiscing about the old, pre-military days with my mom and sisters. The few friends I did have in the Air Force were other outcasts. Our little band that spent time together considered ourselves the Air Force version of the "Island of Misfit Toys." During my work shifts, I stayed invisible. My social life became nonexistent after Chris broke up with me, and being alone turned into real loneliness.

Introverts do better than extroverts when it comes to being alone. As a card-carrying extrovert, I've never met a stranger. I talk to everyone. So being isolated from others made my soul ache. I figured once new people joined my unit at work, the baggage of S.Sgt. Morin would be behind me, and I could start over with new friends. That didn't happen. It was as if someone in my unit held an orientation class where they told new airmen, "That over there in the corner will end your career. You befriend her, and you can kiss your military career goodbye. Don't joke with her, don't talk with her, don't look at her. She doesn't exist."

My dating relationships fared no better. Chris had showed me enough disrespect with his indifference, alcoholism, and cheating to last me a lifetime. Had I stayed with him, it could have turned out even worse.

At work one day, I received a group email sent from our dorm counsel president, SJ Kim. I had seen emails from him before, but I don't think I ever read them. This one, though, featured a subject-line I couldn't ignore.

HUNGRY?

As soon as my eyes hit that word, I clicked on the message.

"Tired of chow hall food? Treat yourself to a special chaplain-sponsored event this Wednesday night at 6:30 p.m. The chaplains will be serving up free spaghetti and pizza! Come empty, leave full!"

Hell, I was depressed, but I wasn't dead.

I asked a friend, "Did you see this email from SJ Kim? Free food. I'm going! You want to come?"

"Oh, Papa Kim?" she asked.

"No, it's SJ Kim," I clarified.

"Yeah, I know his name," she explained, "But we call him Papa since he's old."

"How old are we talking, girl?" I asked.

"Real old," she confirmed. "I'm thinking like twenty-six or something."

Age is relative. Most of us in the dorms were twenty or younger. When someone still lived in the dorms at twenty-six? We considered those dinosaurs.

Then I remembered I had met him once before at a party one of my friends dragged me to in an effort to snap me out of my funk.

"I'm tired of looking at your sad face around here," she had said. "Tonight, we are going to party! You need a change of scenery!"

Her solution? A house party full of wild, uninhibited airmen.

Everything around me took my mind back to Chris, and it didn't put me in a good place. The party noise, like music and drunken laughter, added to my funk. So instead of staying in the center of things, I walked around the house seeking a quieter place. That's when I walked into the kitchen and saw SJ Kim. First impression: CUTE!

When my girlfriend came into the kitchen to get another drink, I pulled her aside.

"Who's that sitting at the table?" I asked quietly. "He's pretty hot!"

"That's SJ Kim," she said. "But I don't know if he's into Black girls."

I got that all the time. *Why say something like, "I don't know if he's into Black girls?" Hell, I know I'm Black. And if he looks over at me, he's going to know I'm Black. It's not like we'd date for a month, and then he would look over at me one day and say, "Whoa, I just noticed something. Are you Black?"*

"Well," I answered, not letting her in on my inner loathing of her comment, "I'm pretty sure he's going to figure out what color I am if he looks over at me. Could you ask him if he's seeing anyone? And maybe have him check me out over here?"

She goes over to SJ Kim, and the two of them talk for a minute. At one point, he looked up, and his eyes found me. Then a big smile crossed his face, and he gave me a little nod. I guess he didn't notice that I was Black. *Thank you, friend, for trying to make me feel insecure.*

"He thinks you're a doll," my friend reported back to me with a pat on the arm.

Since my dating IQ proved low with Chris, and the whole S.Sgt. Morin situation lowered my work IQ, just hearing confirmation that an attractive guy found me cute gave me enough of a boost for one night. My self-esteem lifted from the toilet bowl to the toilet seat. *Progress.*

My friend and I decided to leave the party soon afterwards. Heading back to our place, my friend casually pointed to a vehicle in the parking lot.

"That's Kim's car," she said, pointing to a red BMW Z4. "And that," she said pointing to the next spot over, "is his motorcycle."

For several days afterwards, I kept a lookout in hopes of spotting SJ in his car or on his bike, and I looked for him any time I approached our dorm. But I never saw him. He was like a ghost. Over time, I forgot all about him.

Now, as I still contemplated the invite, I remembered who he was. "Oh," I shouted as I remembered the party where I first saw him, "SJ Kim! Papa Kim! The guy from the party with the sweet rides!"

"Yeah," she confirmed. "That's the one. And I think he was into you."

I wasted no time in responding to his email.

"I'd love to come," I wrote. "Where is it being held?" I played dumb. His email had the address on the bottom: Norfolk Dorm.

SJ wrote back right away, "I'm sorry. I forgot to include that. It's at Portsmouth Dorm."

Glad I asked. I'd have gone to the wrong place.

To make a long story short, I don't know if I screwed it up or he screwed it up, but he sent me to the wrong place! After looking all over the place at the wrong dorm, I gave up and returned to my dorm. And that's when I found the party. And I wasn't happy: they had run out of food.

Temporarily, I forgot about food and went looking for SJ. I found him sitting at a picnic table eating spaghetti and looking good.

"Hey, I hope you're enjoying that nice spaghetti dinner while I'm eating a nice plate of nothing at the wrong dorm where there was no party or food," I told him. "You owe me dinner," I laughed. "They're out of food. The dining facility is closed. And I'm starving!"

He laughed, and it was a good laugh. He had a nice smile, very calm and warm.

Some people around him offered me some of their food. I declined.

"Thank you, but no," I said. "It's his mistake, so it should cost him," I flirted a bit, accepting a piece of pizza SJ handed me from his plate while he laughed and shrugged sheepishly.

The next day, I got an email from SJ.

"Hey, I hope you're not still angry about last night," he wrote.

"I'm good," I wrote back. "It's a good thing you still had some food on your plate, or you might be missing some fingers," I teased him.

"Good," he replied. "Well, how about I make it up to you? Would you like to have dinner with me Saturday night? I'll take you to a nice place, I promise," he offered.

"Okay," I teased him, "but I didn't get a full free dinner like you promised, so I'm not paying."

"I wouldn't dream of it," he wrote. "You won't have to pay. My treat."

And that's how it's done, I proudly thought to myself after snagging a date with a hot guy.

On Saturday night, right on time, SJ knocked on my door to pick me up for dinner. I had never experienced that before. Chris would tell me to meet him downstairs, and if I weren't waiting for him, he'd lay on the horn until I came running out. A couple of times, I hadn't run quickly enough, and he left without me.

When we got to SJ's car (did I mention it was a sweet ride?), he opened the passenger door for me. *Another first.* And on the passenger seat were a dozen red roses. *Damn, I could get used to this*, I thought. He looked like a model, and he showed me such sweetness and chivalry that he had me blown away before we ate.

Is this what a healthy relationship is like? I wondered. *Because I'm not hating this!*

Over dinner, I got my first good look at him away from the distraction of others. Yes, he reminded me of a really good-looking, Korean Mr. Clean. And what's not to like about that? Smooth head, broad smile, and huge muscles. I learned that he had trained as a bodybuilder, and looking at his physique, I had no trouble believing him. His button-up, teal shirt fit nicely over his huge chest and bulging arms. Even his jeans looked fancy, like he'd ironed them. I could tell he was conscientious about his physical appearance. I learned that he took clean eating and exercise very seriously.

"So," I asked between bites, "what does SJ stand for?"

"What do you mean?" he asked, his face showing confusion.

"SJ," I said. "I'm guessing the initials stand for something. So what is your real name? Like Scott Jeffrey? Or maybe Steven Julian?"

"Oh, that," he laughed. "No. But funny story, though," he said as he told me that his mom had not named him SJ at all, but since his mom was from Korea, what came out of her mouth sounded to the nurse like SJ. The nurse asked his mom to repeat the name several times, and his mom kept saying something that sounded like SJ. "SJ," the nurse repeated to his mom. "You're saying SJ." Finally, his mom gave up and shrugged as if to say, "Close enough. Let's just go with that." SJ stuck. And according to SJ, his parents never spoke about what his name was supposed to be.

I enjoyed SJ's company, and I enjoyed the meal equally. I ordered a plate of the greasiest food ever, and I wolfed it down in about five minutes. I did this for two reasons. First, this was a little test. Some dates would say something like, "You're a girl. You should order a salad." Or instead of saying anything, they would just give a look that said, "Damn! You keep eating like that and you'll be five hundred pounds by the time you turn forty!" This allowed me to see if SJ was one of those guys who wanted to put me in a box or expected me to stay tiny into my nineties. My second reason for ordering a high-fat plate of food was, I'm not going to lie, I knew I'd enjoy it. *If you got a problem with me eating what I like, move on*, I'd always thought. *Life's too short to eat lettuce when onion rings are an option.*

"Wow," SJ said with seeming appreciation. "I didn't know you were that hungry." Before I could give him the stink eye, he added, "Would you like to order another plate of something?"

I checked his face carefully. I saw no trace of judgment or sarcasm.

"Seriously," he offered again. "Order something else if you're still hungry. I want you to be happy."

"No," I laughed. "I'm fine. It just tasted so good!"

"How about frozen yogurt to cap off the meal?" he asked.

"It's like you *know* me already!" I laughed. "I'm so up for that."

We didn't make it to froyo. Well, I didn't make it. I fell asleep in his car on the way there. I had the habit of going to bed early, and it was already past my bedtime. Once I had a full stomach, the eyes closed quickly.

The next thing I knew, we pulled up at the dorm.

"What happened?" I said, sitting up and looking around, confused.

"Hahahahaha," he laughed. "You've been asleep for forty-five minutes. We were talking, then I heard what sounded like a cat purring in the seat next to me." Then he laughed again.

He walked me to my door, kissed me gently, and said goodnight.

As soon as I closed the door behind him, I thought to myself, *What a man! And what an absolutely amazing evening. I could get used to being treated like a princess.*

Chapter 11

Leg-Shaving Whore

"You can fix things up, but you can't make them all better.

~ Amy Joy

FORTUNATELY, SJ AND I were both E3s at that time, so we didn't have to worry about breaching the rules against fraternization—which would be a non-officer dating an officer.

After our first date, we sort of fell into each other. We did everything together. He'd send me the sweetest texts. And he'd call me just to say, "Hi, beautiful." We ate together, worked out together, and let our worlds meld together—while we slid deeper in love with one another.

SJ showed me attention, gentleness, and love like no one I'd ever met before. I could see spending the rest of my life with him. I had it bad. And the longer we dated, the more I knew that I had found my Prince Charming and Mr. Right all rolled up into one amazing package. I didn't want it to end!

And that feeling lasted a whole three weeks before it started to change.

"Hey, babe," I said one Friday night. "Why don't we get dressed up and go to Harold's party? It's time you meet my friends."

"Are you sure you want to go out?" he asked.

"Yes, I want to go," I joked. "That's why I asked. So come on. We'll have fun. We don't have to stay long."

"Listen," he said, looking up from his couch. "I don't like you hanging out with those people. They're not good for you. All they do is drink, and they're a bad influence on you."

"But you haven't even met them," I argued. "How do you know they're a bad influence?"

"I don't want you to spend time with them, period," he said to close the discussion.

"Wait," I objected. "These are my friends. They were around before I even met you…"

"Kia, we're in a relationship now," he argued from the couch. "That changes things. When you enter an exclusive relationship," he said, waving his hand between the two of us, "you need to cut out all the people who could hurt that relationship."

"Babe, I don't want to make you jealous," I said, trying to see his viewpoint as I felt my body stiffen slightly. "But these are more than my friends. They're my coworkers, the few that still talk to me. Please let me go this one last time?" I begged.

"Okay, fine," he finally agreed in a huff. "But I'm going with you, so there's no funny business."

"Great," I said, getting changed. "That's all I wanted."

As soon as we entered Harold's place, the first person I recognized was his roommate, James, who rarely attended Harold's parties. James was a civilian. He knew from experience that weird things could happen at these house parties, and he didn't want any drama or trouble. The first time I met James, I was dating Chris, who liked my bubbly, outgoing confidence. On that night, I had introduced myself to James just as he was heading out to the store.

"Great, I'll go with you," I had said, inviting myself.

James had a motorcycle, and I hopped onto the back like I'd known him my whole life. In hindsight, I'm so glad James wasn't a creep. We shopped, returned, and that was that. But ever since that night we met, I felt safe with him, like he was one of the good guys. Which was weird, because most of the guys my age didn't make me feel safe at all.

Anyway, when I saw James at the party, I rushed over to where he sat on the couch.

"Jimmy!" I yelled, giving him a hug. "I haven't seen you in forever. How are you?"

"I'm good, Kia," James replied. "Are you doing okay?"

"I am, Jimmy boy," I smiled. "I'm here with my new boyfriend, SJ," I said as I looked around the room, found SJ, and waved at him.

"That's great. Good for you," James said.

Is it good for me? I wondered for the first time. Because when I waved at SJ, he responded with an icy face.

I returned to SJ's side and introduced him to a few of my friends. I'm not a person who needs alcohol to get out of my shell. I don't have much of a shell to begin with. Most of the time, my personality is like a bottle of champagne after a few shakes. I'm ready to go! Watching SJ, though, I realized that he almost needed to drink to relax. He wasn't a big drinker. I noticed, though, that after one beer, he seemed to loosen up a bit, which was good. But then I noticed after two beers, he started getting paranoid.

"What's wrong, babe?" I asked him.

"Nothing," he said curtly. "Are you ready to go?"

"Well, no," I answered. "Are you?"

"Yeah, let's go," he decided.

"Okay, we can go," I answered, looking around to see if anyone else stood close enough to hear. "Thanks for coming, I guess."

On the drive home, SJ let his tongue fly.

"Those people are no good, do you hear me?" he yelled, as he bounced his hand off the steering wheel. "We're done with them. What a bunch of idiots!"

This "conversation" continued as we walked into the dorm. At first, I let him have his say. But he kept going until I had to fight back.

"Well, I like those people you call idiots," I defended them. "They're my friends, and I was having a good time seeing them."

"I'm glad it's over," he said as he threw himself back down on the couch.

"Well, I want to go back," I said, hovering around the door.

"Kia, you need to decide right now," he said as he got off the couch, crossed to where I stood, and got in my face. "You walk out that door, we're done. It's either them or me. But you can't have both."

I know I looked shocked. He hovered in front of me with his hands on his hips, waiting for my answer. He had dropped an ultimatum on me, one I never saw coming. *Were we even at the same party?* I hadn't seen or heard anything at Harold's that could cause this kind of reaction.

"Okay, fine," I said, tossing my purse on the table. "They're friends. Nothing more. I choose you."

I've often fantasized about how amazing it would be if each of us were born with three do-over moments. Think of a time where you made a mistake. What if you could go back to that instant and make a different choice? How might the course of life take a completely new direction by changing course around a few critical moments.

If I had three do-over moments, one of them would be to return to that conversation and do things differently.

"Kia, you need to decide right now," SJ said to me in that moment. "You walk out that door, we're done. It's either them or me. But you can't have both."

"I know," I'd say. "Goodbye, SJ. Be gone by the time I get back. I don't ever want to see you again."

But we don't have do-over moments. We're stuck with do's, don'ts, and consequences for both. Do-overs are merely wishful thinking.

Later that night, listening to SJ's snoring, I did some thinking. When I'd met SJ, he had such an outgoing spirit. But the longer I knew him, the more introverted he appeared. Early on, we went out to eat all the time. We would walk, laugh, talk, and get lost in one another. As more time passed, he seemed to want me all to himself. As an extrovert, I got my energy by being around other people. Not all the time, of course. But after a long day—much of the time feeling isolated and ostracized at work—I loved spending time with a group of friends, just hanging out.

Are we opposites? I started to wonder. *Was he play-acting the first few weeks when he seemed so kind and open? Or is he just stressed right now and needs time to feel comfortable with my friends?*

I awoke the next morning determined to commit to SJ with my all. *Last night*, I told myself, *was just a blip. Love means compromise*, I figured. *We're in a new relationship, both feeling a bit insecure and carrying our own baggage.* I let go of expecting perfection. After all, I'd never had a normal, healthy relationship before. And since I had no peace at work, SJ felt like my closest friend.

Yeah, I needed peace somewhere. I decided to find it with SJ.

I continued calling and texting my friends, but I did as SJ asked and didn't see them any longer. What I didn't know is that SJ got ahold of my phone, copied names and numbers of my friends, and then reached out to them one-by-one.

"Lose Kia's number, okay?" SJ told them. "Don't ever call, text, or see her again."

"Kia's a big girl," some retorted. "She can make decisions for herself."

"Is that right?" he asked. "Here's a decision I've made. If you ever contact her again, I'll kill you. Is that clear? Or do you want to keep pushing me?"

"Yeah, okay. I got it."

My friends dropped out of my life overnight. It would be a long time before I knew why. While I missed them, SJ and a new cycle of classes kept me so busy, I didn't notice immediately.

In August of 2011, I started summer classes, and the outside temperature felt more like July. Looking through my closet, I picked out an aqua blue sundress with vibrant yellow flowers, my most summery outfit. It flowed down to my ankles, touching the tops of my feet. It had spaghetti straps so I could catch a breeze on a hot day such as today. As I shaved my legs in SJ's bathroom, he came in.

"What are you doing?" he asked.

"I have class today, so I'm just getting ready," I answered without looking up.

"Why are you shaving your legs?" he asked.

"Because I'm wearing a sundress," I responded. "I don't want to look like a sasquatch."

"Why aren't you wearing pants?" he asked without laughing at my joke.

"Maybe you're used to this weather, but it's hot out there for me," I answered, as I put down the razor and slipped into my sundress.

"Who?" he started. "Who are you going to see today?" he asked.

"Well, since it's class, I'm probably going to see the professor and all of the other students," I answered as I finished zipping up my dress and straightening myself in the mirror.

Finally, he asked his *real* question.

"Are you going to cheat on me?" he demanded.

"What?" I whipped my head around. "No! I'm just going to school!"

"Don't give me that bullshit," he yelled. "My ex, that's what she did when she started cheating on me. Shaved her legs and dressed up like a whore. You're going to cheat on me too, you bitch!"

"What the hell?" I asked, wondering how we descended so quickly into cursing, name-calling, and wild accusations. "No, I'm just going to class. This is a sundress. I'm not dressed like a whore. It goes down to the damn floor!"

"Yeah, yeah," he screamed. "Whenever she would go cheat on me, she'd look just like you!"

Then his face flushed bright red and he started crying. In tears, he rushed over to me, grabbed my shoulders, and started shaking me like a ragdoll. I didn't know if he was experiencing a nervous breakdown, had post-traumatic stress, or planned to hurt me.

"Are you fucking kidding me?" he screamed, which told me all I needed to know. I went instantly from confused to scared.

Pushing me away, he ran outside while I watched from the window. Outside the dorm, I watched his body shake. He literally trembled, like someone shivering uncontrollably in the cold or having a seizure. Then, facing the brick wall of our dorm, he punched it several times. I probably imagined it, but it felt like the dorm shook from his blows.

Then he ran back into our room. Once inside, he ran into the bedroom, leaving a blood trail from his knuckles across the floor.

"Oh my god," I rushed after him, trailing the blood stains. "Are you okay?"

Inside his room, he fell to the floor weeping. Torrents fell, leaving streams down his face, and he no longer looked like the threatening man he'd been just seconds before. I watched him, stuck between fear and shock.

"What in the world is wrong with you?" I asked, confused. I let out the breath I had been holding, a little less scared now that he seemed to be out of his fit of rage. Seeing him on the floor, he looked helpless, so I relaxed a bit. "What's going on?"

"I feel like it's happening all over again," he sobbed. "My ex cheated and lied all the time. And I just love you so much," he said between sobs. "I don't want you to leave me."

"Babe, I'm not leaving you," I assured him. "I'm just going to school. That's all."

As he continued to sob and shake his head, I kept trying to find the words that would defuse his raw emotions.

"I'm sorry, SJ," I apologized. "I won't do it again. The clothes I mean. I can change my clothes if you want. But I'm going to be late."

As I started to back away, he grabbed me.

"Promise me you won't do it again," he wailed, grabbing my face with his bloody palms. "Promise me!"

"I promise," I said, trying to end his suffering.

Then he dropped to his knees and started bowing to me. I understood that he meant this as an Asian gesture of humbling himself and asking forgiveness. Then he held me around the waist and hugged me.

"I'm so sorry," he cried as he tried to gain his control. "I'm so sorry. I love you. I don't want you to leave me ever. The shaving and the dress," he said, "it just reminded me so much of my ex. Please forgive me. I'm so sorry."

"It's fine," I assured him. And we stayed like that for a few minutes—me standing while he gripped his arms around my torso, like a toddler clinging to his mother.

Finally, his breathing returned to normal and he stood up, hugging me in the process.

"I'm sorry, SJ, but I'm seriously late for class. Can you give me a ride?" I asked, hoping we could go back to some kind of normal.

"No," he shook his head, looking at me up and down while I sought his eyes for an answer that would put everything right. "I can't drive you to school dressed like a whore. You can walk."

I didn't have time to process what had happened, let alone change if I wanted to make it to class. I grabbed my bag and speed-walked to class. I got to the building thirty minutes late. It wasn't until I stepped into the restroom that I noticed the blood stains on my cheeks. I quickly washed my face in the sink before joining my class, erasing the outward signs of what I'd just endured.

Once I stopped sweating in class, I realized I had zero attention span. My mind kept replaying the tapes of what had just happened at the dorm.

Shaving my legs and wearing a sundress makes me a whore and a bitch? I asked myself. *Then I've been a whore since thirteen.*

The more I thought about it, the more angry—and resolute—I became. I knew I had to walk away.

He had shamed, shook, and offered "I'm sorrys" to me within a matter of minutes. And then he had shamed me again.

"I can't drive you to school dressed like a whore. You can walk," were his last sweet nothings to me as I gathered my things for school.

Screw him, I told myself.

And I didn't just tell myself that, either. Later that day, I told my friend I was done with him.

"He's crazy," I said. "I'm done. I can't do this." I shook my head and wore a face as if I'd just smelled spoiled milk.

"Oh my gosh, no," she said. "What happened?"

"What happened is that he's freakin' nuts," I answered, telling her the entire story, ending with, "And then he says, 'I can't drive you to school dressed like a whore. You can walk.'" I added that I got to class late covered in sweat and blood.

"Not Papa Kim," she said sadly. "Everyone loves him. Maybe he's just having flashbacks, like post-traumatic stress from his ex," she offered.

During breaks, I followed the same friend and another good friend out to the smoke pit, where I reiterated my feelings.

"If everybody loves him," I spat out, "they can have him. Not me. Fool me once," I said and just let the rest of the sentence fade away.

After class, we walked together back to my dorm, and then they followed me upstairs, where I'd planned to grab my things from SJ's room to clear out. But when I opened the door, SJ sat on a chair looking broken. Then I noticed my two friends follow me inside and close the door after themselves.

When I heard the door close, I turned around. My friends just looked at me, and one of them said, "Why don't you just have a seat next to SJ, Kia?"

"What the hell is this?" I asked, like the Queen of Seeing Things Coming.

"Look, Kia," they started. "We wanted to come back with you because we know what happened. You told us, and SJ reached out to us, too. And he's so sorry for what he did," they explained. "He is just such a really, really, really, great guy. And we want you two to work through this, because we believe in you both."

I looked over at SJ, who was sitting crisscross-applesauce with his hands clasped in his lap like a sad, lost puppy.

"He really loves you, Kia," they continued. "You know that, right?"

"You guys know that we've only been dating for like a month, right?" I asked, incredulous that they were doing this to me right now.

"But he didn't mean it," one friend pleaded. "We all make mistakes. Look at him. He loves you *so* much."

"Don't you remember all of the crap Chris put you through?" my other friend added. "He treated you like complete shit. And how many times did you keep coming back for more?"

"Oh, I see," I whipped back. "Since I was a doormat for one asshole, I should just be a doormat for the next one, right?"

"SJ's not an asshole," she answered. "He just made a mistake. And I don't want you to make another one by breaking up with him when you can have something really great with him."

"He told you about his ex, didn't he?" my other friend picked up as soon as the other stopped talking. "He's been through so much. That's all it is. He knows he overreacted, right, SJ?" She now turned to him on the couch instead of me.

"I never meant to hurt you, Kia," he said like a little boy. "I thought I was over what my ex did to me, but I know now that I'm not. I should never have taken it out on you and treated you like anything other than the princess that you are."

That's when I said, "Fuck you, SJ! We're done."

Well, I would have if I had do-overs. But I didn't.

Instead, I sat quietly staring at them all as they kept pleading on SJ's behalf.

"Fine," I said as I caved. "I'm sorry for what you've been through," I said softly. "I didn't know you had so much hurt from your past. We'll get through this together, okay?"

"I promise you, Kia," he said. "This will never happen again. She hurt me so badly, but I will never take it out on you again."

Then he started to cry again.

When a man cries in front of his woman, he's telling the truth and speaking from the heart. But if that man cries in front of his woman's friends, he's even more special.

I decided I didn't care what I wore any longer. I would dress to make him more comfortable, not myself. It was a small price to pay to keep the peace and to help him overcome the bad experiences from him ex.

"I'm not going anywhere, babe," I said to silence him.

And at the time, I meant it.

Chapter 12

Bullets, Blood, and Bombs

"You're not selfish for wanting to be treated well."

~ Jason Momoa

A FTER A WHILE, I found that, like clockwork, SJ and I had a major blowout every four weeks. Years later, someone shared the Power & Control Wheel (go to the website below to learn more) developed by the Domestic Abuse Intervention Project in Duluth, Minnesota, and I had a huge aha moment. [If you or a loved one suspect domestic violence, you can learn more at https://www.thehotline.org/identify-abuse/power-and-control.]

Throughout our relationship, SJ used *intimidation*. Since his arms were the size of my thighs, it didn't take much for him to physically intimidate me. Of course, he also used *emotional abuse*, like calling me "bitch" and telling me I dressed like a "whore." He *isolated* me from all my friends. Whenever we had a problem, he *denied* responsibility and *blamed* me for the conflict. Frequently, he claimed *male privilege* by making all the big decisions for the two of us. Later, he used *economic abuse* by putting me on an allowance not just for my spending-money but for groceries as well. If I made him mad, he withheld money, and I couldn't buy food. Finally, he used *coercion* and *threats* to "keep me in line."

While social workers know all of this, especially those dealing with victims of domestic violence, I figured it was part of the normal ups and downs couples would face and work through together. Social workers would tick off the four stages on their fingers: tension building, incident, reconciliation, and calm.

But I wasn't a social worker. I was a nineteen-year-old girl just coming off an extremely toxic dating relationship. I didn't know any fancy labels, but I knew we fell into a predictable cycle: Bad, Worse, Good, and Peace. You can't make an omelet without breaking a few eggs, right? Well, I couldn't have Good times or Peace times until I first went through the fire of Bad times and Worse times.

Even when we were in the Bad or Worse part of our cycle, I was determined to make things work between us. I needed at least one part of my life that functioned well. And I didn't have that in the office. Some days at work seemed pleasant, and people smiled, waved, and said, "Good morning." The next day, I'd become invisible again, and no one would establish eye contact with me.

So yeah, damnit, I worked for the United States Air Force, not Disneyland. Instead of walking around saying to myself, *All our dreams can come true, if we have the courage to pursue them*, my mantra became, *See nothing, hear nothing, say nothing.*

No matter what I did, though, I'd hear my name whispered around each corner. Some called me Jakia; others called me Kia. Just try to say either of those quietly and see if those K, I, and A sounds used together can possibly be another word. Go ahead and use a rhyming dictionary while I wait here. Okay? So maybe I was paranoid. Or maybe everyone around me suddenly joined the cult of IKEA. But more likely, people were talking about me behind my back. And it wore me down.

In our US military, the second leading cause of death is not war; it's death by suicide. Veterans and active military die by suicide at a rate twice the norm of the general population. Part of the cause is trauma from being in active war zones and seeing terrible things. But the other part of it is the culture of the military. It's like a breeding ground of bullies and hazing that you can't escape.

Military service and prison have something in common: you can't simply walk away. In every other job, you can put in your two weeks' notice and be done. But when you're in the military, you sign a contract. I know that athletes, rock bands, and some corporate executives sign contracts, too, but the cost of their breach of contract is financial. A serious breach of a military contract could result in a dishonorable discharge, or even worse, prison—either of which makes you unemployable for all practical purposes.

I lost count of the number of times I had superior officers tell me, "Just go kill yourself." Unfortunately, too many active military members do just that. When you find yourself surrounded by ongoing negativity, it takes its toll. Eventually, you feel like you're going crazy. I signed up to serve, to do my duty, and I followed orders and directives to the best of my understanding.

I tried to be objective, to take an unbiased view of myself to see what others might witness when they looked at me. Did I put off a vibe? Did I act like I hated my job (which even though I planned on hating, I ended up loving)?

My desire to outrun whatever it was that made others see me as flawed drove me to act like the "early bird" and show up ahead of my shift each day. My mom used to say, "If you're not fifteen minutes early, you're late." I set my alarm and started showing up fifteen minutes early to the office.

One day, my supervisor saw me working in an empty office before anyone else had arrived.

"Hmmph," he grunted. "You think you're setting the world on fire? Back in my day, I used to come to work an hour early."

Willing to do anything to show him how much I wanted to move on from the past, I started arriving an hour early. But I couldn't start working. I just sat there, because the earlier shift still trudged away, taking up each workstation. I made coffee, read through hard copy manuals, and tried to be useful.

That still wasn't good enough. Most days I sat there at work before my supervisor arrived. Once he heard that I came in an hour early for my shift, he asked me to send him a time-stamped email when I got in to let him know I was at work. I did that for years. I didn't get paid to show up early. My promptness didn't win me any grace or credit, either professionally or socially. But I felt like I had to do it for myself, to prove I was willing to do absolutely anything in my power to be the best airman I could be.

More than once, this compulsion of mine created friction between SJ and me.

"Let me take you to lunch before you start work," he offered one afternoon.

"No, I'm sorry. I can't. I need to get ready for my shift," I said as I laid out my clothes for the day.

"What?" he asked, making a face. "It's 12:30. I thought you started today at 3."

"I do," I confirmed. "But I have to shower, get dressed, and beat traffic."

"Okay, but that's almost three hours from now," he objected. "You look like you're ready to hyperventilate!"

"I can't go to lunch, okay?!" I snapped. "By the time we go out, eat, and come back home, I'll be late. I'm already a nervous wreck. I can't let anything mess up my schedule! I don't want them to have any reason to give me shit."

Yet nothing I did mattered. I resigned myself that the harassment would never end.

If you haven't served in the Air Force like I have—as a young, Black female—you might think I'm the exception, or even that I did something to bring this upon myself. Maybe. But no.

Several of us took turns in the doghouse. My only comfort in being hazed at work was knowing that once I finished my turn, the bullies would move on to someone else.

It's my turn now. I'd look at the patterns in office behaviors as if I were studying a zodiac table. *Once I'm done, it will go to Cisco, or maybe Pattersen. Then I can get some work done, maybe even feel good about myself. Oh, I better schedule time to study once my turn is done, since I won't be able to concentrate once they start to gang up on me again.*

A good friend of mine named Heather often became the object of abuse at work. I felt sorry for her, but it was her turn.

One day I walked past her desk and heard her laughing. I stopped to see her.

"What's going on?" I asked. "What are you giggling about over there?"

Heather still laughed when she looked at me, but I saw tears pouring down her face.

"I was just thinking about this dream I had," she said, bursting into laughter again. "It's just so crazy. So Mr. Coates is giving his morning briefing, and everyone is listening. Someone brought in doughnuts. And then I come in through the back door real quiet like. No one sees me, no one hears me. And then BANG BANG BANG! I just start shooting everyone!"

She stopped to wipe more tears from her face. After taking a few deep breaths to regain her composure, she started laughing again—a low, throaty laugh that sounded like it should be coming from an old, chain-smoking man instead of such a young woman.

"Okay," I said, my eyes big and my heart pumping in my own throat.

"Girl, it was crazy," Heather continued between low laughs. "There were bullets flying everywhere. I shot everyone. Then the cops came, but I knew they would come, so I'd already gone out the back door and ran next door where the commander and the rest of the leadership team works. The night before," she paused to laugh and wipe more tears, "I placed bombs all around this building! So while the cops come in to find the active shooter, I'm fixing to blow this one up. You know? I'm going to blow up everybody. And then BOOM! The whole building came down!"

I tried to breathe normally and keep my face calm. But I am one expressive sister. I'm sure my eyes bulged out of my head like I had just seen a ghost as I heard her describe this scene in such gory detail—while laughing all the while.

"Girl, it was so big, bad. It made the news. Everyone was talking about it. It was crazy," she laughed. "So, when I tell you don't come to work, don't come to work."

"Okay, no," I nodded. "When you give me the word, I'll stay home that day. I will."

"My god, you should have seen it!" She laughed and cried some more. "I made a beautiful mess of everyone. Blood everywhere. Screams. Fire. It was so crazy."

"Wow," I responded, always articulate in times of crisis. "That was one crazy dream," I nodded, not wanting to disagree with her on any level. "You know what, though?" I said apologetically. "I got to use the bathroom real quick. You stay right here, okay? I'll be right back."

She grabbed my arm and pulled me towards her face as I moved.

"Don't tell anybody, okay? Let this be our little secret," she said softly.

"Not a word," I promised. "This stays with us."

In truth, I did have to use the restroom, because watching and listening to my friend had scared the shit out of me. But the bathroom could wait. I made a beeline to the office of my flight chief.

"Sir," I started and told him everything as quickly as I could. "I don't know if she's joking or not, but I think she needs help."

"There's nothing wrong with that girl," he shook his head and started to go back to work.

"No, sir, seriously," I pleaded. "She needs help. I think there's something wrong with her, like mentally. I'm telling you she laughed like a demon and cried at the same time. Heather's like my best friend here, but I think she's snapped. She's going to hate me, because I told her I wouldn't say anything, but she needs help. If I don't say anything, people are going to die. I don't mean to snitch on her, and I know how people have treated me, and they treat her the same way, and…"

"Alright," he said, holding up his hand for me to stop talking. "You're always into something, aren't you? You're always saying something when you should be minding yourself."

"Sir, I'm sorry," I tried one more time. "But I think she needs help."

"Just leave," he said with annoyance.

I didn't know if he would do anything, but it broke my heart to see my friend in such agony. I had to say something even if it led to me being mocked even more.

Eventually, he talked with her. After about fifteen minutes, I watched the two of them walk out together. I didn't see her again for more than a month. She didn't answer her phone or respond to any of my texts.

Then one day, Heather returned to work like nothing had happened. As soon as she set her things on her desk, she came over to my desk.

Before I could apologize or raise my hands to protect my face, she hugged me.

"Thank you so much, sweet Kia," she said while she held me. "Thank you so much for saying something. I really needed time away from this office. I needed help more than I knew. I was going to lose it. I was going to blow this place up."

We sat for a moment before she kept sharing.

"This place. I hate it," she shook her head. "It made me want to kill myself. But then I started thinking, 'Why would I kill myself? What I should do is kill everyone else.' I hate this place so bad, but it's better than where they sent me."

"Where'd they send you? What was it like?" I wanted to know.

"Some of the people down there are crazy," she said. "I mean, not crazy like me. I'm talking apes-throwing-shit crazy. My roommate would pee in the corner of our room. That's when I knew there are all different levels of crazy. I'm *here* crazy," she raised her hand to her chest level. "But some of them are *here* crazy," and she lifted her hand above her head.

One of my best friends had cracked, and I saw it happening to her. We had all arrived at Langley AFB so happy after Basic and tech school. But day after day and month after month of feeling worthless tore us down. We had been there for less than two years, yet we felt old, used up, and broken down.

Chapter 13

Conflicting Opinions

"If you don't know what you deserve, you'll always settle for less."

~ Rob Liano

A S A CHILD, I remember reading that if you put a frog in a pot of boiling water, it will jump out immediately. However, if you put a frog in warm water and continually increase the heat, that same frog will relax in the warmth until it's eventually boiled to death. I have no idea what book I read this in, or why in the world anyone would want to try to boil a frog to death in the first place. But that story stuck in my mind.

It didn't take me long to figure out that in my entire relationship with SJ, I was that frog. When we first met, he felt warm and soothing. He made me comfortable. Over time though, he heated up to the point of boiling over. Yet I kept staying, believing I could again experience his warmth without suffering the burns. What I felt in our courtship and marriage was a cycle of simmering and boiling. At some point, I accepted that I would one day be found boiled to death.

One day I had lunch with a group of friends from the office. As I said, they were hot and cold. But whenever coworkers showed me signs of friendship, I readily latched on with gratitude. We were at a restaurant ordering food when SJ texted to find out where I was. I let him know the name of the restaurant. He texted back, "Did I say you could go to lunch?" I explained that the entire office was going, so I felt like I needed to be there too.

"I'm coming over," he texted.

He rode his motorcycle to where we were eating. Since we had already ordered, I asked him if he wanted to get something, too.

He made no response except a cold glare, which he kept on his face the entire time we were there. He didn't order food. He ate nothing. Just stared.

After lunch, one of my coworkers offered me a ride back to work.

"No, but thanks," I said as cheerfully as I could. "SJ's got an extra helmet. I'm going to catch a ride back with him."

I climbed onto the back of his bike, and he took off. Oh my God. He accelerated quickly and put on the brakes the same way, and I was holding on for my life. We finally got on the highway, and I watched as he sped up to 105 mph. He started bobbing and weaving between semi-trucks and cars, literally not giving two shits. Tears of fear started streaming down my face.

If you've ever been on a motorcycle, you know that you lean as you turn. If you don't lean into turns, you'll fall off and crash. Each time SJ turned and leaned, I had to lean in unison with him. At one point, my knee was so close to the ground that I could feel pebbles kick up against my leg.

I started to pray. "Hey God, it's me again. I know we haven't talked for a while, but if you get me through this moment, I promise I'll be good. If you just let me live, I'll go to church more. I'll do anything you want. Please just let me live!"

SJ made another quick stop when he pulled up in front of my office, slowing just enough for me to jump off quickly before he took off again, leaving rubber on the pavement.

I knew after work that I'd be going home to a verbal beating, but I still didn't know what I had done to upset him. As I predicted, as soon as I got home, the yelling started. He didn't wait until we got into either of our rooms. Instead, he started screaming at me in the parking lot as others were getting off work and pulling up.

"How the hell can you embarrass me like that?" he demanded. "So now you're just going to take off and not tell me when you have lunch plans?" Then he grabbed his helmet and took off, leaving me standing in the parking lot looking like a rebellious teenager who'd just been scolded by her dad.

I went back to my room and just waited. I couldn't go anywhere without him. I mean that literally. My car became unreliable, so I sold it. Then he talked me out of replacing it, saying we didn't need two cars and a motorcycle. Except his car was a manual, and he never taught me to drive it without making the car jerk like a bucking bronco.

This became the norm. Every two weeks there was something new. Then after he raged for a while, he'd come back and apologize, adding another excuse for his misbehavior.

"My parents didn't love me," he told me. "They left for Busan, South Korea, during my senior year in high school, and they never came back, not even for graduation. They didn't even tell me they were leaving. I came back from a senior trip, and they'd left me a note telling me they'd decided to move. Nothing I did was ever good enough."

And every two weeks, I ended up comforting him like he was the victim. I took on the role of his soul healer—his surrogate mother figure—believing that the reason he treated me that way had to do with the love he never got as a child.

"I'm so sorry, SJ," I consoled him. "I don't know why they did that either. I wish they could see the man you are today."

I didn't know the words for it at the time, but I was in an abusive and codependent relationship. I put his needs above my own needs—and safety. Why did I stay? I wanted to fix him, to give him what his parents hadn't, and to show him that *there is good in the world*. I wanted to *be* the good in his world.

And there was another reason, one a little more personal and selfish: SJ didn't judge me for what happened to me at work. He supported me and often agreed with me.

"It's not you, Kia," he'd tell me over and over. "Some people are just assholes, just bad people. You need to fight them. You are a strong woman. Do not let them get to you."

SJ became my rock when things were bad at work. I had no one else to hold onto. My other close friend experienced a breakdown due to the work stress. I didn't want to add to her stress, because I had seen how her own stress had taken her to a very dark place. Just like SJ was the only person in my life, I was the only person in SJ's. He didn't have any friends, nor did I. We seemed stuck together.

One day I got a Facebook friend request.

"Who is this guy sending you a request?" SJ demanded.

"I don't know who you're talking about," I answered honestly. "I haven't been on my computer."

I didn't know it at the time, but SJ kept login information to all my social media accounts so he could monitor everything I did.

Looking over his shoulder, I saw the name.

"Oh, that's a coworker," I responded. "I hadn't seen that request."

"Well, how do you know him?" he persisted. "Why is he friend requesting you? How much have you talked to him, to where he feels comfortable adding you on Facebook?"

"I have no idea. We work together. We talk at work," I responded, not yet sensing that this conversation had taken a turn.

"Do you go over to his house?" I sensed things getting worse. "Have you slept with him?" he finally asked bluntly.

"Okay, um, no," I snapped at him. "And I don't have a car, so how would I get to his place? You drive me to and from work every day."

"So you're saying you would sleep with him if you had a way to get to his place?" he continued.

"What the hell, SJ?!" I screamed. "How do you go from me getting a friend request on Facebook to the only reason I'm not sleeping with him is because I don't have a car? I'm not sleeping with him. I don't want to sleep with him. I wouldn't sleep with him if I had a car, a boat, a helicopter, or a space shuttle. Do you have any idea how crazy you sound?"

After I exploded, he started crying again and repeating how his ex-wife cheated on him, how she had been called the "dorm mattress" because she had these so-called friends everywhere that were nothing more than *friends with benefits*. Since his ex had cheated on him, he felt certain that I would stray, too.

His tales of woe always had their desired effect on me. I would reassure him and forgive him.

And the cycle of abuse continued.

Our *up* part of the cycle was to get away to Abingdon, Virginia, where we stumbled upon a beautiful little area on the water. Abingdon became my happy place. Nothing bad ever happened when we went down to Abingdon. Whenever we fought, I knew that afterwards we would go to Abingdon. Those were the good times of our dating relationship, sort of like how Huck Finn and Jim from the Mark Twain book *The Adventures of Tom Sawyer* felt when they were on the river in their happy place. In Abingdon, we were free from conflict.

But just like Huck and Jim couldn't spend their entire lives on the river, SJ and I couldn't stay in Abingdon forever. Returning to the real world usually resulted in us returning to a dark place shortly after we returned to base.

For Christmas, we decided to head over to Ohio to join my family for the holidays. Not surprisingly, SJ and I had a huge blowup shortly before we started the 500-mile drive. Nothing like being trapped in a small car with an irritated, angry man.

After we'd been home for a couple of days, my sister Brandy caught me alone.

"SJ, huh?" she said as a statement more than a question.

"Yeah? So?" I responded to the question part of her statement.

"I don't like him, Kia," she told me flat-out. "My gut's never wrong. Something is not right about him."

"What do you mean?" I responded. "You just met him."

"It'll come to me," she nodded. "I can't put my finger on it just yet, but something about him is not right."

My whole life I wanted nothing more than approval from my sisters. I didn't get that when I was little. But by the time I joined the Air Force, we had grown into a tight family. I wanted Brandy to like SJ, because her approval meant the world to me.

Not only did Brandy not approve of SJ, but she also issued a warning.

"If you keep seeing him, it will not end well," she said. "Mark my words."

This would be a great moment for a do-over opportunity. I should have listened to my sister and ended things with SJ. My sister was right, and even more annoyingly, I knew she was right even then. But still I held on.

Without SJ, I'd have no one.

Shortly after we returned to base, my stomach started to fall out of me, and I fell sick. SJ brought me to the hospital and called my family, who came to Virginia to see me. My insides hurt all over, and I thought I must be dying.

My sister Alexis stood by my bedside holding my hand.

"Listen, Kia," she said with a face full of sisterly love and compassion. "If it turns out that your kidneys are failing, I wanted to be tested as a donor. I want you to have one of my kidneys."

"Oh my God!" I said to her with panic in my voice, my eyes wide. "What have you heard? Have the doctors told you that my kidneys are shutting down? Do I need a transplant?"

"No, no," she said with a smile. "They haven't said anything about that. But I want you to know, you can have one of my kidneys."

As tears started to fill my eyes over such a selfless, generous offer of a gift, my sister continued.

"Because I hate my job," she winked, "and I could really use a few weeks off work."

I laughed as much as my weak body would allow. I loved (and love) her to death.

A few friends stopped by the hospital to visit as well. While they raised my spirits, their presence angered SJ.

"Why are these guys coming to see you?" he demanded once a couple of male coworkers left. "They don't need to be here."

"They're my friends, SJ," I answered. "I'm so glad some of them are being kind and supportive. That's all it is. You know I don't have many friends at work, so see this as a good thing."

While I was writing this book, a male friend reached out to me to let me know that SJ had called him after he came to see me in the hospital. "Stay away from her," SJ had said, threatening my friend.

A military doctor ran multiple tests on me. While awaiting the results, he came into my room to see me.

"A physical exam shows nothing unusual, nothing to explain your pain or vomiting," he said. "We're looking for the source, but we haven't found anything yet. I just have to ask, are you sure that you're not…"

"What? Pregnant?" I asked, horrified.

"No, not pregnant," he shook his head. "We tested that immediately. No, what I want to ask is, are you sure you're not doing this to yourself? Are you making yourself throw up?"

Since then, I've had several female friends tell me their own stories about male doctors suggesting that they are hypochondriacs or making themselves sick whenever they can't find an immediate cause. It's like, unless a woman comes to the hospital with a sword sticking out of her skull, they assume the symptoms are fake or stress-related.

"No, I'm not making myself throw up," I said as calmly as I could. "You pushed all around my guts, and I told you where it hurts. And your nurses have watched me puke. Ask them if they've seen me doing this to myself."

The test results proved two things: first, I wasn't crazy, and second, the military doctor was a jerk. Well, the test didn't document that part about him being a jerk, but I drew that conclusion when the results showed I had a stomach virus and kidney infection. Another doctor started me on antibiotics and another medication. In a couple of days, I felt good enough to be released back to work.

Although my illness hadn't been psychosomatic, I have no doubt that stress played a part in it. Fights with SJ were ongoing. The court-martial of S.Sgt. Morin approached. Just thinking about testifying put a knot in my gut. The harassment at work remained relentless. I've heard that trauma can weaken the immune system. I don't know if that's accurate, but my body started responding to the stress around me in ways I hadn't experienced before in my life.

A few days after I got out of the hospital, my mom called me.

"Let me tell you something, Kia," Mom said. "That SJ? He's a keeper."

"Oh, that's nice to hear, Mom," I said, wondering what he had done to win her over.

"When you were in the hospital, he watched over you like a mama bear looking out for her cubs," she told me. "He was so responsive to you and your every need. Such a sweetie, that one. He's a keeper, that's for sure."

I didn't fully recognize it at the time, but I could have taken the opinion of my sister and the polar opposite opinion of my mom to divine the truth. If my sister says,

"Something ain't right," trust her. She's incredibly intuitive. If my mom says, "He's a keeper," that should have been the kiss of death.

Chapter 14

A Whirlwind Quasi-Romance

"Marriage has no guarantees. If that's what you're looking for, go live with a car battery."

~ Emma Bombeck

WHILE MOM FELL in love with SJ, I still wanted some freedom. He had a controlling nature, and although I told myself he treated me that way because he loved me and his parents had screwed him up, I had reservations about a long-term relationship.

As Valentine's Day approached, I got excited.

"I can't wait for Valentine's Day!" I told him while we watched TV one night.

"Valentine's Day is so cliché," he said. "It's nothing but a Hallmark holiday. I don't need one specific day to show the person I love that I love them."

Well, what are you doing with the other 364 days? Because you damn sure don't show me love every day. I'd be happy if you just went a few weeks without yelling, I thought to myself.

"I just don't think our love should be confined to one day," he concluded.

Since he hated Valentine's Day, I had to hate Valentine's Day, too. Except I didn't. Inside, I wanted to be loved, celebrated, and maybe a little worshipped. But I'd be okay with just being loved. From then on, I stopped looking forward to that day, and I told myself, *It's just another day.* And I'd tell my friends, "Yeah, me and my man, we don't need one day to say that we love each other. He treats me amazing throughout the entire year."

Deep down, though, I really wanted to celebrate Valentine's Day, even if it was just a pretend holiday. When you can't have something, you convince yourself that you don't want it, and you might even become critical of those who have it. You lie to yourself, so you don't feel the hurt. But you still feel it. You just get used to disappointment.

A month after a Valentine's Day of nothing, he proposed to me.

Leading up to that "big" moment, he was on his best behavior.

"Honey," he told me one evening after work, "I want you to know that I'm trying hard to show you how much I love you. I know I don't always get it right."

"Thanks, babe," I replied. "I know you're really trying. It shows. I love you, too."

Then he started crying.

"It's just…" he broke down. "I'm so stressed out. I could never be good enough for my parents, even though I've tried so hard to get their approval. My dad is old and sick. Do you know what it's like to have parents that old?"

His dad was born in 1926.

"And then being Asian," he cried out. "It's the worst. They put so much pressure on me to become a doctor or lawyer! But I told them I wanted to serve my country by joining the military, and they shook their heads at me. My mom said, 'Tell me when you're an officer. Maybe then you'll have something to be proud of.' They put so much pressure on me!"

While my own career was going down the drain, SJ's star was rising. I caught a little draft from SJ's reputation at work. SJ had the respect of everyone. Even the wing commander knew SJ personally. Aligning myself with SJ took some of the negativity off me and might even help me rehabilitate my military career, I knew. And, of course, despite my misgivings, I loved him.

While I slept on the evening of March 22, 2012, SJ slipped a ring on my finger. Call me the Queen of the Clueless. That ring sat on my finger for about three hours before I noticed. Going into the bathroom to do my business, I put my hands on my waistband when the light pinged off something on my hand.

"What the hell!?" I shouted and ran into the bedroom.

SJ stood in the bedroom. "So," he said expectantly, "what's your answer?"

"To what?" I answered, always quick on the uptake.

"The ring?" he pointed to my hand.

"Huh?" I asked, not fully understanding what was happening.

"Will you marry me?" he finally asked.

"Oh my gosh, yes," I gushed. "Sure! But wait, wait. That's not how you do it! You have to get down on one knee. Why aren't you down on one knee?"

"Okay, okay," he said as he dropped to his knee. "Jakia Monee Clark, will you marry me?"

"Yes! This is so sweet!" I told him as we hugged.

My friends were like, "Tell me again how he proposed."

"He's very romantic," I told them. "I work swing shift, and he works day shift. It's like our schedules never match up. He was so sweet, and he proposed to me in the middle of the night. I got to wake up to this amazing surprise!"

I tried to spice that story up as best as I could. If I could have worked in that he got an entire orchestra of elephants playing "Wind Beneath My Wings," I would have said that, because I felt like the true version needed some hot sauce or something.

I'm a bit ashamed to admit it, but I felt disappointed by his proposal at first. It began when he nixed Valentine's Day. I could understand his sentiments on that, even though I would have liked to have been swept off my feet for at least one Valentine's Day in my life. And now this, a marriage proposal with no fireworks, billboards, or marching bands. Even flowers or candy would have meant the world to me. All my silly little girl dreams were disappearing. I told myself it was okay, that I didn't need big shows of love around my proposal—even though I'd eaten alone at Arby's with the same amount of romance as this big moment.

Then I reminded myself that SJ did what he could given our crazy schedules, and that made it special. Our story fit our crazy lives. I never thought I would get married, but now I was. I lowered my expectations by embracing my story to retrofit my emotions. I redefined those foolish, childhood ideas of romance and happy-ever-after.

What we have is raw, I told myself, *but it's real.*

A few hours later, SJ said, "So let's get married!"

"Whoa," I laughed. "We've been engaged about five minutes. What's the rush?"

"Well, we can at least pick a date!" he added.

"I'm thinking next year, maybe in the springtime," I told him.

"That's one thought," he nodded. "I was thinking more like now."

"How can we do it now?" I questioned.

"All we need is a marriage license and a courthouse," he explained. "And witnesses, I guess. It's easy. "

"But then my family can't join us," I argued. "I really need my family to be there for my wedding day. It would kill them if we got married without them being invited."

I didn't mention that it would kill me, too. I was so homesick. When times were hard at work, I leaned on SJ. When things were hard at work and with SJ, I leaned on my family for support or at least a sounding board. *How could I get married without my family supporting me?*

"Look, getting married at the courthouse is paperwork," he explained. "Just a legal formality. Let's get legally married now, and then we can do a real wedding and ceremony later with everyone. I'm going on deployment soon. We need to be married so if anything were to happen to me, I could provide for you even after I'm gone."

"But get married twice?" I asked, still trying to wrap my head around it.

"We need to be legally married," he explained, "which is just simple courthouse ritual. Then I can add you to my insurance and make you my beneficiary if I get killed in action. We'll do a real wedding once I get back, and we'll have your entire family be part of it, I promise."

Ignoring my preference, I let him convince me. I didn't want him to go on deployment feeling insecure about our relationship. We opened our calendars. The soonest available date we were both free turned out to be April 11. Engaged March 22, married three weeks later. Every girl's dream, right?

"Okay," I conceded. "But did you ask my mom? Did you ask my dad? You got to do the right thing by asking them for my hand in marriage."

SJ called my mom, and she said, "As far as I'm concerned, you two are already married!"

SJ and Mom viewed marriage as a formality. I did not. I wanted some part of our union to follow a traditional path. "Little things" like getting down on one knee and talking with the parents first were important to me. I could let go of Valentine's Day, a splashy proposal, and a fancy wedding. But some things I wouldn't budge on.

SJ had no relationship with my biological father, There Daddy. SJ hadn't even met or spoken to him on the phone. So I called There Daddy myself.

"What the hell are you thinking?" he asked over the phone, sounding pissed. "I haven't even met this man. How long have you known him? A week? A month?"

SJ wasn't the only one of us with daddy issues. My dad had been more like a distant relative most of my life. I'd see him from time to time when we visited Tennessee. I'd learned at my Air Force graduation ceremony that long ago, my parents had a misunderstanding that led to me having little regular contact with my dad. The two of us had just started communicating after the ceremony, around a year prior to SJ's proposal.

"I shouldn't be critical of a man I don't know," my dad settled down a bit. "But it sounds like you're rushing into things here. Baby girl, just slow down. He's not like milk, that if you keep it too long it goes bad. Just slow the hell down, baby."

We argued back and forth, with me adding brilliant statements like, "What do you know?"

"Baby, all I'm saying is there's something wrong with a man who wants to marry you so quickly," he concluded.

And since I was nineteen years old and knew everything, I countered with, "There's nothing wrong with him. He's perfect."

Yeah, SJ wasn't perfect, but I'd be damned if I were going to let anyone else badmouth him!

The conversation with my dad didn't end well. Following that call, we went a long time without speaking to one another.

On the morning of April 11, I put on my most conservative dress and wore a shawl over my shoulders. Even though I'd told my dad "I know what I'm doing" and "I'm a big girl and can make my own decisions," part of me feared I was making a huge mistake.

Between the dorm and the courthouse must have been twenty stop lights. I remember praying as we approached each one, *Please turn red! Please turn red!* I wanted more time, to slow things down, for God to give me some sort of sign to run screaming. But God gave me nothing but green lights the entire drive. Next, I prayed that we wouldn't find a parking spot. Yeah, the parking lot was as empty as if it were a federal holiday.

As we walked inside, sweat spotted my dress. Not from the heat, because it was still cold outside. There may have been patches of grey slush from winter snows still in the shadows. My nerves jangled. I went to the bathroom and threw up. *That's normal, right?* I coached myself. *They call these pre-wedding jitters.*

Yet I kept praying that something would go wrong to delay the wedding. Like maybe the marriage license expired, the judge got food poisoning, or a meteorite hit the courthouse. Something. Anything!

Before we went in front of the judge, something besides my stomach *did* come up—but the sign made me even more determined to marry SJ. Members of my leadership team, one of them my section chief—the one that issued me the letter of counseling for texting S.Sgt. Morin—showed up along with one of my best friends. As the three of them entered the courthouse, the leaders pulled SJ aside as my friend invited me into the corridor.

"What's going on?" I asked her.

"The hell if I know," she said. "They told me that they were taking me out to lunch. The next thing I know, we pull up to the courthouse. Now I learn you're getting married? Dang, girl! Well, congratulations!"

"Yeah, thanks," I said, confused as hell.

Inside, my leadership team pleaded with SJ.

"You're making a huge mistake," they told him. "She's trouble. Don't marry her. She is going to ruin your career. You can go places, but she's going to drag you down with her."

When I returned to the courtroom, I heard just enough to put it all together. She was a distraction, someone they brought in to occupy my attention so they could stop the wedding. Before I could say or do anything, SJ threw them out.

"You aren't welcome here," he said calmly. "And I don't ever want to hear you say anything like that about my wife again."

Hearing that, my butterflies and fear disappeared. I knew what my leaders thought about me. In their eyes, I was a troublemaker and a rat. They had told me as much to my face. My supervisor had told me weeks before, "If you're not happy here, feel free to kill yourself."

I'm a good person, I reminded myself as I straightened my shoulders, put my arms around SJ, and kissed him like I had the first time we kissed. *I am not shit. I deserve to be married.*

And with that, the judge motioned us forward. A couple of *I do's* later, and we became husband and wife legally.

Chapter 15

How You Know the Honeymoon Is Over

"What counts in making a happy marriage is not so much how compatible you are but how you deal with incompatibility."

~ Leo Tolstoy

S J DIDN'T GET me flowers or a mylar balloon to announce his love for me on Valentine's Day. Instead, we ordered Chinese takeout that night and watched *Law & Order*. His proposal lacked fireworks and a violinist courting me with a love song. But he got me a ring, and he got down on one knee when I asked him. He didn't throw me a traditional wedding like I'd pictured since I was a child. *But so what? I told myself. In the end, we still got married. And now for the big honeymoon celebration!*

Which we had back in our new apartment later that day. Yeah, one thing SJ had going for him after our first few dates was, he consistently underwhelmed in the romance department.

You've heard the phrase "the honeymoon is over" to describe the time when the euphoria of dating meets with the reality of marriage? I thought people used that as a bitter joke, and I never imagined that it would be a thing in my marriage. But our honeymoon didn't last twenty-four hours before our marriage reached a boiling point.

SJ and I sat on the couch watching TV when he started talking about his friends and his prior life.

"They sound like really great people," I said. "I'd love to meet them."

"Yeah, sure," he said as he pulled up messages on his phone.

"So this guy here," he said as he went through different messages and explained how he knew various people. He shared interesting stories about each person, so I could get a sense for how they fit into his life. Finally, he handed me his phone. "You can get an idea of how they talk. Like that guy here, he's hilarious! He always pretends to be a stupid redneck."

As I read each message, I felt like I was catching some insight into SJ. In the Air Force, SJ earned the respect of leadership. And why not? He worked hard to project a "can do" image and great professionalism. Though I'd first seen SJ at a social gathering, he turned reclusive once we became a couple, and he didn't have any close friends as far as I could tell. But reading messages from his friends, I loved seeing how he'd built and maintained relationships with friends from high school and college, even ones he hadn't seen for years.

As I kept scrolling, I saw a familiar name, someone I recognized as one of his exes. I started to scroll past her message, but then I saw the date: December 24, just a few months ago. *He was messaging her while we were visiting my family over Christmas?*

Curiosity with just a little jealousy crept in, and I opened the message. What I read nearly burned my eyes: "I love you, and I want no one else but you. I've never met anyone that compares to you. You're the one I've always wanted. You will always be my number one. Please get back with me."

That shit read like a Frankenstein version of all the worst love songs ever written.

"SJ?" I said as I held the message to his face. "Why would you send her something like this when we were seeing my family? We were exclusive by then. What the hell is this?"

"Damnit," he said. "Give me back my fucking phone."

"So, was I just your plan B just because she wouldn't take you back?" I demanded. "You married me because you couldn't find anyone else stupid enough to have you?"

He shoved me away from him. "You shouldn't have been going through my phone in the first place."

Things exploded quickly. He started yelling nonsense, and finally I raised my hands.

"Oh, no," I said, standing up and grabbing my purse. "I can't even do this! You're mad at me because you had an affair of the heart with another woman just a few months ago. Screw you! I'm leaving."

Just days before we had moved into this new apartment, and I didn't know the neighborhood at all. But I needed to get out of this place. My head felt like it was going to pop, and my heart ached.

Not having a car, I walked away as fast as my legs would take me. I covered miles while crying so hard my head hurt and my tears ran dry. I called my mom, needing some consolation, unconditional love, and guidance.

After telling her about the fight, I ended by saying, "I'm so lost, Momma."

"Relationships can feel that way," she agreed.

"No, Mom," I clarified. "I mean I'm actually lost. I have no idea where I am. I don't recognize any streets or landmarks. I'm in a part of town I've never seen before."

I tried to describe my surroundings to her as I best could. Mom told me to stay put, and she promised she would get me some help.

That "help" turned out to be SJ.

"My baby is lost," Mom told him. "I don't know what happened, and I don't care. But I want you to go out there and find her," she told him.

"No," SJ replied. "She's the one that took off. She can find her own way home."

"I don't care if she left," Mom told him. "She's your wife. Get off your ass and go find her!"

"If I go get her, she is going to be mad at me," SJ deflected.

"And if she ends up hurt or killed," Mom explained like she was talking to a child, "I'm going to be mad at you. And SJ," she added, "I'm not someone you want mad at you. Go find her."

My Knight in Shining Armor wouldn't budge.

"No," he said before ending the call. "She got herself lost. She can get herself found."

I walked until I ended up on a campus, but I didn't recognize anything. I kept walking a grid up and down streets, hoping something would look familiar. Finally, I found a place to eat. I was glad I'd had my wits about me enough to grab my purse as I left. Since I was running on empty, I sat down to eat and rest my feet. When I finished eating, I pulled out a small work notebook I had in my purse. I wrote out my thoughts, hoping that were I to review them later, I could figure out where I'd gone wrong.

Then I started walking again, eventually finding some familiar sights. After roaming the streets for nearly nine hours, I found my way back to my apartment.

Before I walked through the door, I pictured what would play out when I stepped inside. SJ, I imagined, would greet me with a sincere apology, broken for his betrayal and wracked with worry for my absence. He'd then beg me for forgiveness and promises that I was his one and only true love.

That's not what happened. Instead, SJ sat on the couch stone-faced, his cold eyes glaring at me with disgust.

"When you're ready to apologize for invading my privacy, I will listen," he said callously. "Until then, I have nothing to say to you." Then he left the room.

I felt like I had just entered *Seinfeld's* "Bizarro World" where everything was the opposite of what it should be. Hell, I was only too familiar with that at work. A supervisor sleeps with my boyfriend, and threatens to have me thrown out of the military if I open my mouth. Another guy says he wants to bend me over the desk and do me from behind, and I'm told that I'm too sensitive. Another pervert charges me with his dick out, and I'm the villain for reporting him. *But here at home, I need to apologize for finding out I was my husband's plan B from looking at his phone after he handed it to me? I'm the bad guy here?*

Later, I called my mom to let her know that I made it home okay—and to cry on her shoulder.

"I made such a big mistake marrying him," I wailed. "I want to file for divorce. Everything about him is a lie. He was basically cheating on me, Mom! I can't trust him ever again. I don't think he ever wanted me in the first place. I was just his placeholder until his ex took him back."

"Stop being a child," she reprimanded me. "You made your bed, now go lie in it. Grow up, honey. You're a married woman now. Act like it."

My mom called me back for hours, but I didn't get her calls. Once she got off work, she sent me an email.

> "Let me start off by telling you that I love you dearly…But Jakia, don't you ever walk out on your marriage or husband…. It gets a little hot, and off you go. Not only worrying SJ but worrying me, your mother. I understand that you were angry, but that did not give you the right to worry me like you did. You walked off without so much as a plan as to what you were going to do—childish—and you had **BLATANT DISREGARD** for your own safety—**VERY CHILDISH**! I was so sick I was literally throwing up.
>
> "You are EFFIN married! This is not a boyfriend/girlfriend situation where you can just break up and leave without a second thought or a second glance back. You will be twenty years old tomorrow! So please **GROW THE FUCK UP** and act like a grown married woman. If the two of you don't get together soon, you will be in divorce court before the ink is even dry on your marriage license. **WORK THIS SHIT OUT!** And don't think SJ is off the hook. His email is on the way."

I can't say I received my mom's words with an open heart or mind at the time. I was pissed! But then, I knew where my mom was coming from. From my perspective, my mom had always put men on a pedestal, especially when she considered that man godly. *A good woman needs to defer to her godly man.* This is no dig on my mom. It's just a commentary of what most women of her generation and upbringing believed.

For example, one time I told my mom I wanted to get my hair cut short. She didn't tell me how she thought that style would look on me. Instead, she asked, "Did you ask SJ? You need to ask your husband for permission. He may not like short hair." From Mom's point of view, a wife needed to be subservient, cook, clean, and get behind her man.

SJ knew this, and he took advantage of it when it suited his needs. Whenever the two of us had conflict, he would get my mom involved, relying on her to reel me back into the fold when I strayed. I could hear him talking to my mom on the phone from the next room.

"Talk to your daughter," SJ would say. "She's at it again. She's not listening. Here's what happened…"

Moments later, Mom would call me.

"Jakia, what's going on?" she'd ask. "SJ just called me crying. Why can't you listen to him?"

"Mom," I'd argue, "he's manipulating you. I heard what he told you, and that's not what happened at all. Let me set you straight on what really happened…"

I swear, when SJ and I argued, he had my mom on speed dial!

Putting my mom in the middle of our struggles weakened my relationship with her. Instead of my mom listening to me and offering support, it seemed like every other time I spoke to her, she would lecture me about how I'd been mistreating SJ.

In less than a year, SJ isolated me from friends—and then my own mother. At the time, I figured I'd been so busy with the court-martial of S.Sgt. Morin that I'd lost people for being a bad friend. But even years later, I'd hear from former friends how SJ had threatened them if they didn't stay out of my life. Mom had become my only lifeline, and when SJ rallied her to his side, I had no one.

Unfortunately, our lives don't stay in tidy compartments. Shit spills over from one part of our lives into the next. When my home life and work life both went in the crapper, I was alone with my misery.

I still had a hangover from the court-martial of Morin, the man who exposed himself to me and then came at me. Just months earlier, I pleaded again with my commander to drop the charges, or at least to allow me to skip testifying; but I learned I had no choice. My first sergeant informed me that I gave up my Fifth Amendment rights when I joined the military. Then she told me that the commander had personally selected the individuals that would attend the hearing. I had been the key witness. Months later, I still had nightmares from the trauma of S.Sgt. Morin dropping his drawers in front of me and the part I had to play in his court-martial.

I did what I had to do, and I answered every question put to me on the stand. As difficult as it was to relive those events, in the back of my mind I held onto the hope that once the ruling was finished, the worst would be over.

S.Sgt. Morin was found guilty. As a result, he received months of confinement, was stripped of his rank and benefits, and received a bad conduct discharge. He was out of the military.

But the worst wasn't over for me.

"You all happy with yourself?" a coworker asked me days later. "You got a good man punished for nothing. He's got a wife and a new baby at home, and now he's locked up and unemployed. Nice job, Penis Girl. And you wonder why no one wants to talk to you and you get treated the way you do. It's because you're a fucking nark!"

In February, I got a new supervisor, S.Sgt. Edwards. I hoped that a new boss would mean a new start for me—and an end to the verbal abuse. During my first meeting with him, he set the tone for what I could expect from him.

"I was in the room to hear your case. From where I sit, I think you both should have been kicked out. I knew Sergeant Morin. I considered him a friend. And I just want you to know, your time with me as your supervisor is going to be hell," he informed me with ice in his eyes.

I vacillated between trying to get help by speaking up, and accepting my fate by shutting up. But when I walked out of Edwards' office, I felt I had nothing to lose by reaching out to his boss.

"Not everyone is going to like their supervisor," his boss said as I nodded in agreement. "But you have a top-notch NCO in Edwards. He's in the limelight since his return from deployment and the Purple Heart he received. Look at it this way. He's the poster child of our organization. And you? You are in a bad light due to the Morin court-martial. You have no credibility, your image is shot, and everyone will believe an NCO over an airman any day."

He confirmed my worst fears: *I was screwed.*

Colonel Hicks, my group commander, held a townhall shortly after the court-martial to share the details and outcome of the case. From a training perspective, I understand why he did it. By putting the case out in the open, he was showing that the Air Force had zero tolerance for the kind of misconduct S.Sgt. Morin displayed. But that message got lost somewhere in translation. Many people in the room grumbled that "he got ratted out" and it was a shame for that to happen to a "godly, Christian man."

Later that same month, my unit held a Sexual Assault Prevention and Response (SAPR) session where the details of my case were again shared. But this time, the entire group was encouraged to discuss how the situation could have been avoided.

"What did the airman do that was wrong here?" the facilitator asked the group assembled for the mandatory meeting. "Do you agree with the verdict? If not, what do you think should have happened?"

He asked these questions as if no one knew who the "airman" was. All eyes turned to me, many of them rolling in contempt, as if my "misbehavior" had necessitated this mandatory meeting.

Airmen responded from around the room.

"She should have had her ass kicked out, too," one airman said while looking directly at me, receiving a chorus of nods from others.

"She was stupid," another said. "How could she not know what was going on? She played dumb because she loved the attention."

More nods and shouts of agreement.

I stood up, starting to leave the room, but I was told to sit back down. In a room full of airmen, NCOs, and officers, they all seemed to agree that the case had been mishandled, since I had not been court-martialed, too. My face flushed hot, and I bit my lip to keep from crying.

The male airmen in the room were the most vocal and critical. Since they made up three-quarters of the attendees, they made a lot of noise. Most of the female airmen either looked down or tried to keep their faces neutral. But one airman, the woman sitting next to me, reached over and took my hand.

"I was assaulted, too," she whispered. "This is bullshit."

Were it not for her support, I would have lost it right there.

"That's bullshit!" she yelled out. "So now we court-martial victims?"

"Shut up," airmen yelled throughout the room. A few "boos" echoed off the walls.

She sat back down and burst into tears. Now it was my turn to comfort her.

To show you how my work and personal life overlapped, do you remember what I said about my new supervisor, S.Sgt. Edwards, the one who told me, "Your time with me as your supervisor is going to be hell"? Months after becoming my supervisor, it was Edwards who showed up at the courthouse to talk SJ out of marrying me.

No, there's not retaliation in the military. Who could ever think such a thing?

After the required training program, my coworkers felt empowered to harass me with impunity. The few newly-arrived airmen to our unit who didn't know S.Sgt. Morin and had no history with the situation succumbed to the peer pressure to demean me for the crimes others attributed to me.

Once word got around that I had married SJ, ugly comments slathered in racist remarks spread through the office.

"They'll be divorced in no time," another said. "We should start a pool! That would be awesome!"

"I heard she got knocked up. That's the only reason they got married," I overheard one saying about me.

"Maybe we should get her a book. Do you think she's already got a copy of *What to Expect When You're Expecting a Nigger-Chink Baby?*" another added.

Can you imagine any workplace in the country that would tolerate an employee saying that? I can. The United States Air Force. For all its zero-tolerance posturing, the Air Force tolerated flagrant harassment, and members of leadership even took part in it at times.

Fortunately, SJ and I moved from the Worst part of our cycle—into the Good part. As an overture for peace, SJ suggested a motorcycle trip to Abingdon, our happy-place oasis. As it always happened in Abingdon, we melted into the surroundings and had a short, positive, loving time together.

When we returned home after our trip, I thought again about what my mom had said. I decided that "growing the fuck up" meant I needed to lower expectations for daily bliss and raise my tolerance for frustration. I concluded that forgiveness meant being willing to move forward even while still hurting.

As I armed myself with those thoughts, SJ and I experienced a month-long period of peace after Abingdon. I started thinking that I could handle this marriage thing if we continued having periods of peace like this.

Sadly, that peace wouldn't last.

Chapter 16

Guard Dog and Girlfriend

"Evil is unspectacular, and always human, and shares our bed and eats at our table."

~ W.H. Auden

IN MAY, SJ and I had a terrible fight, the cause of which I've long since forgotten. It's like trying to keep a rolling count of the number of white cars you see in a year. After a while, you ask yourself, "Why did I even start doing this?" It's pointless.

Anyway, during this argument, SJ got so mad that his face looked like a cherry tomato. I mean, it got *red!* He screamed, cursed, and waved his hands like he was fighting off a swarm of bees. While he boiled over in rage, I had a momentary memory of him walking me out to his car during our first date, opening the passenger door, and surprising me with a dozen roses. Then I pictured his charming smile and heard his soft voice say, "I love seeing you enjoy yourself." And then my memory faded, and I snapped back to reality, where I was the object of wrath from this spit-spewing, red-faced man with arms flailing close to my face.

And I lost it.

"You bamboozled me," I said, shaking my head in disgust. "You *absofuckinglutely* swindled my ass! When we first met, you were this kind, sweet, nice man full of chivalry and warmth. You were the kind of man I'd always dreamed of marrying. And now look! We've been married for two months, and you're this. THIS!" I shouted. "Look at yourself! You are the opposite of the man you pretended to be!"

That's when his head exploded.

"I'm not the fucking man you married, huh?" he screamed back. And then as if to prove he wasn't that man at all, he turned and started punching holes in the walls. "I'm not that man, huh? Is that what you said?" he screamed as he continued to punch the walls.

I'd never seen him so pissed off. Hell, I'd never seen anyone that mad. SJ started shaking so much, I worried that he was going to have a stroke. As he entered this new, heightened state of anger, I went from mouthy to scared shitless. I ran across the apartment to the bathroom, where I locked myself inside.

Instantly, he started pounding on the door, and it rocked each time he hit it.

"Please, stop!" I yelled. "You're scaring me! Please leave me alone! I don't like it when you get like this!"

"Open the fucking door now," he said in a growling tone. "If you don't open the door, it's just going to get worse."

"Stop it, please!" I screamed back.

He stormed away from the door, and I took some deep breaths to try to calm down, and I saw that my hands were trembling. *Just breathe, girl,* I tried to slow myself down.

What I didn't know at the time was that SJ had gone into the other bathroom, where he found a bobby pin. Then he bit down on it to flatten it out, chipping his tooth in the process. He ran back to my bathroom, and he used the flattened bobby pin to try to pop open the lock.

On a side note, one way you know you live in a top-notch apartment is when you can use a bobby pin to pick the locks.

I stood in the bathroom with my arms wrapped tightly around me in a state of panic, and then I heard the doorknob jiggle. As I heard him fumbling with the lock, I began praying as fast as I could that he couldn't get inside the bathroom. Before I could finish my prayer, the door suddenly swung open. I didn't have time to process what was happening, SJ charged with all of his physical force and anger. He grabbed me—like a linebacker would greet a running back—and shoved my body into the side of the bathtub before pinning me against the wall. My chest grew tight, and I could hear the pounding of my heart in my ears. I remember thinking that this moment could not get any worse. Then he started prying my mouth open. I refused to let him, as I clenched my jaw tighter.

"You made me chip my tooth, you bitch!" he screamed while he dug his thick knuckle into my soft cheek. "You're going to feel the same pain I felt." He pressed harder—twisting and grinding as if he was jacking a car—until I had to open my mouth or risk my jaw cracking. Once I released my jaw, he stuck his large fingers in my mouth and yanked against my front teeth. I started gagging, and clawing at his hands, as saliva poured down the sides of my mouth.

"Does that hurt?!" he screamed. "DOES IT?!"

I continued clawing at his hands, I could feel my jaw begin to tire out as I try my hardest to bite down. I didn't know if humans possessed the ability to pull another person's teeth out with their bare hands, but I was not willing to test that theory.

I never imagined that this is what my life would become.

The last thing I will ever see is the image of the man I married taking my life. This is how my life will end, I thought as I gagged.

I couldn't breathe with his face smashed against mine and his fingers in my mouth. He grabbed and shook my teeth, making my head rattle from side to side.

I've seen two different ways people rage. Some take quite a bit of pushing before they lose it. As they get more worked up, they raise their voices, get sarcastic, and toss out increasingly cutting remarks. If they get pushed over the edge, they stay angry for a long, long time. Others have short fuses, and they blow up with hot, sudden intensity. Fortunately, their rage doesn't last long.

SJ was the latter. At some point, he caught a glimpse of himself in the mirror, and he took his hands out of my mouth. Then he walked out of the room without saying a word.

Never before had he put a hand on me. Normally he'd hit himself or the wall. But having his hands on me—in my mouth, while he seemed to enter a blackout of rage— terrified me. I remained frozen in place for about thirty minutes, afraid to move, letting soft wails escape my throat as tears soaked my face.

SJ's shout from the next room broke me out of my paralysis.

"Dinner's not going to cook itself!" he yelled loudly.

I still couldn't will myself to move. Then SJ stomped his foot or hit the wall, causing me to stand up and walk into the kitchen, still dazed.

Over an otherwise silent dinner, SJ apologized for "what happened earlier." He added that he knew he had a problem with rage, and he promised to go to anger management once he returned from deployment. I forgave him. But with each episode, I had less forgiveness to offer. It was part of our cycle of abuse, and I feared it would never end. Instead of believing it would ever feel normal, I waited for the cycle to return to better days.

My new short-term plan was to hold on until he left in June. *Then,* I told myself, *I'll have time to figure things out. Just hold on, girl!*

Before deploying, SJ had some business to attend to in Florida, so we decided to combine business with pleasure and take a mini-vacation.

"I want to make sure you're protected when I'm gone," he told me once we got to Florida. "Let's get a dog."

I love dogs, so I was thrilled and accepted his offer gladly.

I asked him if we could go to the shelter to adopt a dog. He agreed. Once we got there, he chose a corgi as my guard dog. A damn corgi! It looked like a quadruple-amputee German shepherd. That dog was too lazy to guard anything except his own food dish, and even if he were to become aggressive, he'd only be a threat to a little person or a kibble. But I loved him, and he became my closest friend.

Before SJ deployed, he told me that he would take care of all the bills while he was out of the country. His solution wasn't to use online banking or automatic withdraws from an account. Instead, he pre-wrote six checks for each of the bills: rent, groceries, utilities, cable, etc. After we were married, he'd talked me into adding him to my banking accounts. We shared all accounts, except for a banking account that he used exclusively for business, and he told me that he'd be taking the checkbook with him.

"Hold up," I said when he told me his plan. "I had this checking account before I even met you. You insisted that I add you to my account. Now you want to take the checkbook with you? That's not right."

"We're a family now," he explained calmly. "We need to budget as a family. We can't just splurge on things because we have money in our account."

"I agree," I said, conceding on sticking to a budget. But then I pushed back, saying, "But we're talking rent, utilities, and groceries. Those aren't luxuries. They're necessities. And what about if something comes up at the last minute? How is it any different if you write the checks before you go, or I write them as payment comes due?"

"Exactly," he said. "Since it makes no difference, I'll take care of it. And to make sure neither of us spends money foolishly, I'll hold onto the checkbook. You won't have access to it, and I'll be in a country where my checks are useless. And if there's ever an emergency, take it from your *allowance*, and I'll pay you back. Besides, there won't be any emergencies."

Allowance? I'll get an allowance instead of a paycheck? I thought. *Welcome back to being ten years old again. If I brush my teeth without being asked, can I stay up late tonight, too?* I wondered.

I wouldn't call SJ book smart, but his skills in manipulation—especially when he used them in juxtaposition to my desire to trust him—won out. He convinced me that his intentions were to make my life easier, and I believed it—because I wanted to believe that he meant his actions as kindness and not controlling.

After he deployed, I learned the challenge of grocery shopping with a pre-written check. If I went over, I had to take money out of the "allowance" he gave me— money meant for the things he deemed as non-essentials like getting my hair done, buying clothes, eating out, etc. And of course, I'd have to apologize if I went over, telling him it wouldn't happen again, then cut back the next time I shopped. Then he asked me to take a picture of my grocery receipts so he could see what I purchased.

Damn, I thought. *Is he making sure I'm not buying a booty-call kit with strawberries, whipped cream, chocolate sauce, champagne, and edible underwear?*

But I had more peace than I had known for a long time, and I didn't mind being accountable to him in those ways. I mean, I thought it was weird and over-the-top, but if it made him feel better and kept us in a good place, it was a small price to pay.

While he was gone, I spent a lot of time volunteering with junior ROTC kids. In the process, I made a new friend, Vickie. After we'd worked together a few times, she came up with a great idea.

"Hey, the Fourth of July is coming up," she said. "We should have a barbecue! You want to?"

"Oh, yeah!" I said, more excited to have a friend than anything else, considering how lonely I had been.

On my drive home, though, I had an *oh shit* moment: *How am I going to tell SJ that I want to have a friend over?*

Okay, yes, I get how pathetic that sounds. I had been this free-spirited, fun-loving, extroverted child before I met SJ. And I knew I had changed into this submissive, permission-seeking wife. But then I asked myself: *Well, how did things work out with Chris? Chris treated you like crap. So yes, SJ is controlling, and he can be mean and violent. He goes into rages, and is scary at times. But…*

I really struggled to find something to add after that *but.*

When I thought it all through, I knew how abnormal it was to seek my husband's approval to have a female friend over when it could not possibly put him out, thousands of miles away. Crazy or not, I spent a week trying to come up with how to bring up the subject to him.

By the time of our scheduled Skype call, I was as ready as I could ever be. Once we did the usual catching up, I steered the conversation to the upcoming holiday.

"So," I threw out there, "the Fourth of July is next week. I was thinking of doing a barbecue."

"Okay," he said with the inflection of a question full of suspicion.

"But here's the thing," I started. "I met a friend, Vickie. She's a really great girl."

"I know her," he said flatly. "I can't stand her. She's a whore."

"No," I argued. "I heard that, too! But she's like one of the boys. That's why she has so many male friends. And I really like her. I don't think you know how lonely it is being here by myself. I have no one to talk to, and it's starting to lead me into depression."

"Okay, then I can call you more often," he offered.

"SJ," I pushed back, "you're on the other side of the world. Our schedules are completely opposite of one another. I could really use a local friend to be here for me."

Finally, SJ allowed that he would *think about it*. But before he would decide, he wanted to talk with her first to "make sure she's a good person to be around."

Here's the thing: even though I sought SJ's permission to come over for the holiday weekend, Vickie was sitting at my kitchen table while SJ and I spoke.

"Okay, great, let me see if she's home, and I'll get right back to you," I answered.

Then, running into the kitchen I talked to Vickie.

"SJ wants to meet you," I said.

"Sure, no problem," Vickie replied.

"No, not *meet* you," I clarified. "He wants to interview you. Like grill you, to make sure you're a good girl, good influence, good person, all of that."

"Say what?" Vickie answered.

"It's okay," I reassured her. "He worries about me when he's not here," I lied. The truth was, he *controlled* me—when he was here or far away.

Minutes later, I reconnected with SJ and let him and Vickie get to know each other. I had once seen Vickie chug a frosty mug of beer in less than four seconds. I know, because we timed her. She had other charms as well. Like she was, well, *charming*. In no time, SJ gave his blessing.

"You two have fun," SJ said with a smile and a wave. "Do what you gotta do, and be safe."

Moments later, he sent me a text telling me that I could have Vickie over to the house twice a month, but I needed to let him know when she would be there, for how many hours, and what we'd be doing.

"Sure, sure," I texted back, ecstatic that I had his permission to have a female friend over.

In no time, I made up for lost time. It had been so long since I'd had a close friend! Vickie and I held movie marathons, ate popcorn out of a massive bowl, stayed up late laughing and talking, went out to eat, got mani-pedis.

Except for SJ's tight control on money, I felt like a normal woman. If I stayed within my budget, life was good. SJ did, though, put a few other restrictions on my cash. If I overspent on groceries or argued with him, he would cut back my allowance. Still, financial servitude seemed easier than having him home with me.

It wasn't long before I used Vickie as a sounding board about my relationship with SJ. I told her about our first huge fight, the time I dared shave my legs and wore a sundress to class. A look came over her face like she'd just smelled spoiled milk.

"I'm not done," I said. "Let me tell you the rest."

Once I started talking, I couldn't stop myself. I told her everything that happened in our relationship, including my allowance reductions for disagreeing with him.

"Girl," Vickie said after a long pause. "That's the most whacked shit I've ever heard."

"Yeah," I nodded. "But it is what it is."

"No fucking way, sister," she countered. "It is what you've *allowed it to become*. Girl, you deserve so much better. You need to leave him, and leave him now before his crazy ass gets back!"

"I can't," I protested.

"What do you mean, you *can't*?" she pushed. "He's overseas! What's he going to do? Take away your next birthday?"

This was the second time I had allowed the word *divorce* to flash in my head, but this time, it was like a neon sign, a beacon to me.

"Oh my God, you're right," I agreed. "I can sneak away and get divorced while he's gone. Why didn't I think of that before?"

So that's what I determined to do.

Chapter 17

"Just When I Thought

I Was Out…"

"It takes a lot to shatter a person whose soul whispers, 'I've been through worse.'"

~ Lidia Longorio

LATER THAT DAY, I sent SJ a message from my phone over Skype telling him I wanted a divorce. He was thousands of miles away, where he couldn't hurt me physically.

He didn't get my message right away, but when he finally saw it, the notifications from Skype blew up my phone. Due to the time difference, he reached out to me in the middle of the night while I slept soundly. When I awoke and answered his video call, I heard sobbing—accompanied by screams. He was hurt—and furious.

"I want to kill myself!" he wailed. "I'm deployed, fighting for you and our country, and you do this to me?!" he screamed.

SJ had deployed to Al Udeid in the middle of Qatar. If that means nothing to you, picture Florida in an overseas locale. On earlier calls, he told me how much fun he was having at this huge mall where he raced go-carts indoors. He also rode camels, drank beer, and essentially lived in a tropical paradise. I'm thinking: *We rushed to get married in case something happened to you when you deployed. Like what? Heat stroke from too much salt on your margarita glass?*

When he continued to play the I'm-probably-going-to-die-here-anyway card, I just listened.

"I'm in the armory surrounded with guns. I'm going to take one here and kill myself right now while you watch," he said.

I snapped out of smart-ass wife to terrified wife instantly.

"Okay, SJ," I pleaded. "Don't do that. I didn't mean it. I'm not going to leave you. I'm sorry I said that. I'm just lonely, and I miss you. I promise we are going to make things work. I'll do better with communication. You'll see."

Just because I wanted out of our marriage didn't mean I wanted him to suffer or die. I wanted freedom, yes, but not at all costs.

Later that day, I started to question myself. *Was it wrong that I so hated pre-written checks and an allowance from my husband when I worked my ass off full-time, too? Was it wrong that I didn't want to live with constant violence, screaming, and abuse?*

I made a pros and cons list of marriage versus divorce. My cons list grew long; the most difficult ones I faced were his controlling nature and frequent violent outbursts.

But then I told myself: *For better or for worse. I made this bed, as lumpy and miserable as it may be. I need to lie in it.*

After his suicide threat, I messaged him a few times to make sure he was still okay and not in a bad place.

"I'm fine," he responded. "Quit texting me about it."

"A couple of days ago, you put a gun to your head," I reminded him.

"You forget that ever happened," he replied. "I'm not in the same place. And don't say a word to anyone. I was afraid I'd lost you, so I lost it for a minute."

Ironically, around the same time, I started contemplating suicide. While driving home from work after pulling the 11 p.m. to 7 a.m. shift, my supervisor, S.Sgt. Edwards, called me to return to work and see him in his office.

When I got into his office, he closed the door and told me that SJ had learned of *my adultery.* He spent the next ninety minutes tearing into me for *cheating on my husband while he was on deployment.* Then he planted a seed in my mind.

"I don't know how you live with yourself, you whore!" he screamed. "If you had any decency, you'd kill yourself for what you did to him."

No, you didn't miss the part where I wrote about my adultery. Because it never happened! I had never cheated on SJ, so I had no idea where this accusation and attack came from. I hadn't spoken to SJ for several days, since he'd been under radio silence. If I understood my supervisor correctly, SJ had reached out to his own commander, telling him that I'd cheated. SJ's commander then informed my boss about what I allegedly did. And now my boss dropped down on me like a piano from the sky for doing such a low thing while my husband was away.

Repeatedly, I told my supervisor that I had no idea what he was talking about. I gave up trying to figure out how a rumor like this even started, and I became more worried about SJ's mental state. When I told him I wanted a divorce, he threatened to harm himself. I could only imagine how much he would be suffering if he believed I'd cheated on him!

This situation did me in.

As exhausted as I was from the long shift and the after-shift verbal beating I'd just taken, I'd had it. *Everyone thinks I'm a whore, and I should go kill myself.*

Why would such an idea enter my head? Because moments earlier, my supervisor said this to me: "You should kill yourself. Then I can reach out to your husband and let him know how soon he can collect life insurance on you."

My thoughts grew increasingly darker as I drove home. According to what SJ thought of me, I'd failed as a wife. And based on how the Air Force leadership viewed me, I'd failed as an airman. Those parting words from my supervisor, spotlighting these two major failures, did me in. When I could finally see the lake that I passed each day going to and returning from work, I knew what had to be done. I decided to steer the car into the water and let it sink to the bottom with me inside.

But as soon as I pointed my car towards the lake, I thought about my dog at home. I could picture the poor little guy sitting at the door doing the potty dance because he'd been inside for so long. And I knew he'd be hungry, too.

No matter how miserable I felt, I didn't want the dog to suffer because of yet another failure on my part. I changed my plan. I decided I'd go home, take my dog out, and leave him plenty of food and water. Then I'd come back to the lake.

Depression is a strange thing. By the time I took care of the dog, my depression was stronger than ever, but it left me wiped out. I became too tired to move. I ended up sleeping on the couch for ten hours, and my last thought was, *I'm so depressed, I don't even have the energy to kill myself. You really are a loser, airman.*

By the next day, I decided that I would at least try talking to SJ so I could die defending my fidelity to him. I couldn't get through until three days later. When we finally connected, he'd read through the many messages I'd sent him when I panicked about what he'd heard and how he must feel.

"What's going on?" he said, sounding scared.

"I just need you to know, I did not cheat on you!" I told him, sobbing. "I don't know why you think I did, but I promise you that I would never do that to you."

"What are you even talking about, Kia?" he demanded.

I explained what I knew, or at least what my supervisor told me.

SJ blew a gasket.

"I have no idea what you're talking about," he said. "I've never heard anything. We've had no comm link for eight days. Things heated up over here, and we're just getting back online now. But I will find out what happened."

This was more than a simple miscommunication. What S.Sgt. Edwards did was designed to create drama. He'd told me to kill myself due to the shame I'd brought to myself and my husband, and I nearly did.

I later learned where this chaos started and who started it. When SJ's female commander learned this was his first deployment and that he had a new wife at home, she called my section chief asking him to conduct a wellness check. She wanted him to see if I needed help mowing the lawn, fixing things around the apartment, and things like that. My section chief, in turn, asked S.Sgt. Edwards to check up on me. All these were wonderful and kind gestures. But S.Sgt. Edwards decided to take this as an opportunity to torment me. Yes, it was personal. And yes, it almost led me to take my own life.

It didn't dawn on me until much later how when others tried to create a rift between the two of us, we grew closer. By the time SJ returned home from deployment a month later, he and I had patched things up as best as we could. To celebrate his return, I planned a special welcome home for him. Especially after the crap S.Sgt. Edwards put me through by alleging I'd had an affair, I actually missed SJ. I asked Vickie to help me do something special for his return.

The day SJ came home, I took him out to the movies—on my allowance, I might add. When the movie ended, I went into the bathroom and texted Vickie to put everything in motion. I had purchased candles, sushi, and wine for her to bring to my apartment on cue. And then I'd instructed her to fill the bathtub with bubble bath, champagne, and rose petals.

Once she finished setting up everything, she hid in a communal utility closet in the basement until we entered the apartment. Then she slipped out, unnoticed.

"Oh my God!" SJ exclaimed when he saw everything laid out. "What's all this?"

"Surprise!" I yelled and hugged him.

"How did you set this up?" he asked.

"Vickie," I answered. "She worked with me as my silent, invisible helper."

"You let someone else in our house?" he asked, his tone turning dark.

"Yes," I said, nodding. "Did you see the rose petals leading to the dinner table? Did you notice our favorite song playing?"

But I had ruined the special night. I'd *foolishly* let someone else in our house while we were away. No matter what I did, I always did one thing that stuck out to him like a giant nose pimple on prom night. I didn't have the same pain in the pit of my stomach or tightness in my chest I'd had the first few times he'd let me know I'd failed him. But it still hurt. I had begun to accept the fact that nothing I did would ever be good enough.

He didn't like the way I did dishes, so he forbade me from washing them. He didn't like how I did laundry (I didn't know there was a wrong way to use a washing machine, but apparently, I'd found it), so he took that away from me, too. I know, right? I should have shown incompetence in cooking, cleaning, and everything else at home, and I would have had lots of free time! I had no objection to having chores removed from me, but it wore me thin repeatedly hearing about my incompetence.

As part of our dysfunctional cycle of abuse, SJ made it up to me—not with words, but with a wonderful two-week vacation to Busan. This port served as an Asian version of Abingdon for us. We enjoyed one another, and we reconnected away from the stress of work.

Upon our return from vacation, we jumped back into work. SJ volunteered to work front gate duty, but as he got ready to go, he couldn't find his winter gloves.

"You can wear mine," I offered. "When you drop me off at work, I can grab them off my desk for you to wear."

When I got to my desk on that first day after vacation, SJ came with me and stood near the door waiting so I could hand off my gloves. But I found my desk empty. As I looked around, I saw coworkers from across the office looking at me, gauging my reaction. My cubicle walls that had worn Justin Bieber wallpaper had been stripped bare, and all my personal items like photos were gone. I heard snickering around me.

Jerks, I thought to myself. *You're never going to stop messing with me, are you?*

"Are these black gloves that were on the desk yours?" a coworker asked innocently, interrupting my thoughts. At the time, it didn't strike me as odd that she would ask me about those gloves when all of my belongings had been removed from my work area.

"Yes," I answered quickly, standing in shock as I looked at my empty desk and desk drawers.

"Okay, good," she replied. "I packed up your area. I'll just leave them here on top of this open box." She pointed to the back of the office.

"Babe," I said as I motioned for SJ to follow me to the back of the office. Once he joined me, I pointed to the pile of boxes containing my belongings, "You can grab my gloves. They're on top of that box."

He took a few steps across the room to the box, and then he picked them up, jamming one hand into the first glove. He let out a loud cry.

"Shit!" he yelled. "What the hell?!"

He pulled out his hand. Thumbtacks had pierced his hand in several places when he jammed his hand into the glove to make it fit his larger hands. Drops of blood quickly puddled on the floor.

"Oh my gosh!" I shrieked. "I didn't know those were in there. Those thumbtacks were meant for me," I explained, afraid he'd think I had done something like that to him on purpose. Then I showed him that while we'd been away, someone had packed all my belongings into boxes.

"Look," I pointed out things around my office. "They tore down my wallpaper, broke my picture frame—and even ripped apart my Bible."

My Bible! I saw the torn-out pages in the trash, and then they'd tossed the empty spine into one of the boxes.

SJ swore up and down. "They put these *in your gloves?*" he said, pointing to the thumbtacks now spread across the floor. "Are you fucking kidding me?! Where's the flight chief?"

"Please, don't make a big deal of it," I said quietly. "They'll only make my life more miserable if I say anything."

"My hand's fucking bleeding. Look at it! What do you mean, *Don't make a big deal?!*" he shouted.

He briskly walked the office searching for my flight chief while I trailed behind, but he found my section chief T.Sgt. Emil instead.

"Who do you think did it?" T. Sgt. Emil asked SJ.

SJ told him the name of the woman I suspected—the one who had taken credit for moving my stuff and putting my gloves in the box!

They "investigated." All that meant is they looked up the woman's past disciplinary record and found she had no previous action for harassment or hazing.

"Do you have any proof it was her?" T. Sgt. Emil asked.

"When I was out of the office, she packed up all my belongs," I explained. "I mean, everything down to the last paperclip. There's nothing left of mine in my cubicle. That's harassment right there. When I returned, she held up those gloves and asked if they were mine. I said, 'Yes.' I watched her put them in the open box near the exit so I could find them easily."

"Maybe the thumbtacks fell inside the glove when she put them in the box," T. Sgt. Emil suggested.

SJ saw firsthand how my leaders went out of their way to protect people intent on harassing me.

"You could drag a glove through a desk full of loose thumbtacks. Maybe one would stick in the glove. But fifteen? That's how many I pulled out of my hand."

"I'll talk to her," T. Sgt. Emil offered. "But you know, there's no proof she did it, and she has no history of harassment. So there's nothing we can do unless she does something like this again and we have evidence."

"Unbelievable," SJ shook his head.

"Listen," T. Sgt. Emil said. "Your wife has had a few bad things happen in this unit. She's going to be transferring soon to a new section where she can get a fresh start. I think that will put an end to anything you think might be going on here."

I moved out of that unit and started my new job in January. Things improved briefly at work. The same couldn't be said about life at home.

Growing up in the Asian culture, SJ had this thing called *saving face*. Essentially, saving face is an overarching principle to live by where you never do anything to bring shame upon your family. SJ could be screaming at me, threatening to choke me out, but if someone showed up at the door, POOF! The enraged, cursing man disappeared, and this gracious host took his place.

"Please come in," he'd say. "Do you want a beer or something? If you're hungry, Kia can whip up something for you to eat in no time."

He would do anything to keep up this façade, until the door shut. Then, he'd return to screaming as if he'd never paused.

While I say saving face is part of the Asian culture, I already understood it from the Black community. Many members of the Black community share the prohibition against "putting the *bizness* out on Front Street" or "putting folks on blasts." No one wants to expose acts that could bring collective shame to the community.

During this time, I mastered putting on my own face whenever I stepped out of the house. I didn't know it at the time, but I had something known as smiling depression. No matter how miserable I felt on the inside, I would answer, "Great. Everything is wonderful," to anyone who asked me how I was doing. I never wanted anyone to see my tears underneath my smile.

Yet at the same time, it got harder to face every day

Images

Background

The appellant and Airman First Class (A1C) JMC were both assigned to the ███ Supply Chain Operations Squadron, ████████████████████████ The appellant was a staff sergeant with over 9 years of active duty service, and A1C JMC was a 19-year-old, first-term Airman with less than a year on active duty.

A1C JMC first met the appellant when she arrived at ████████ They were assigned to the same duty section and their cubicles were next to each other. The appellant introduced himself to A1C JMC when she first arrived and invited her to his church. Over time, the appellant began making inappropriate comments to A1C JMC. At physical fitness training, he told her and another Airman that they would look good mud wrestling together. He also told A1C JMC that she reminded him of a certain movie star that was sexy and with whom he wanted to have sex. In addition to the verbal comments, the appellant sent A1C JMC e-mails in which he discussed giving her hugs and wrote he would love to run his fingers over her abs. In response, A1C JMC told the appellant he had "crossed the boundaries" and that he had a wife and a child on the way.

1 – "Background"

116

DEPARTMENT OF THE AIR FORCE
█████ SUPPLY CHAIN OPERATIONS SQUADRON █████

MEMORANDUM FOR SRA JAKIA M.█████████   23 July 2013

FROM: SSgt █████████████

SUBJECT: Letter of Counseling

1. On 22 July you texted TSgt █████ that you were unable to fly back into the local area, due to an "emergency landing" on the runway; which canceled all flights. When instructed to seek alternative options to find a way home, you stated "they re-booked my flight for the next day and I will be back." This resulted in having to extend your leave to accommodate you. This selfish act not only impacted the mission, but caused your co-workers to run your assigned reports. Weighing all options to rent a car and drive back to the local area show that you lack the Air Force Core Values.

2. You are hereby counseled. You have failed to follow not only my basic standards and expectations you are in direct violation of Article 92, Failure to Obey an Order or Regulation and Article 91, disobeying a Non-Comissioned Officer. I expect you to follow all rules, regulations, and read AFI 1-1 and understand the BASIC Air Force Instruction and adhere to the Air Force Core Values of Integrity, Service Before Self and Excellence in all we do. If you fail to maintain standards and obey the orders given to you by myself or any others in the chain of command, further administrative action will happen. You have the ability to to perform your duties at and above the standards passed down to you and I look forward to seeing you return to acceptable standards.

3. The following information required by the Privacy Act is provided. **AUTHORITY:** 10 U.S.C. 8013. **PURPOSE:** To obtain any comments or documents you desire to submit (on a voluntary basis) for consideration concerning this action. **ROUTINE USES:** Provides you an opportunity to submit comments or documents for consideration. If provided, the comments and documents you submit become a part of the action. **DISCLOSURE:** Your written acknowledgment of receipt and signature are mandatory. Any other comments or documents you provide are voluntary.

4. You will acknowledge and return this letter to me immediately. You have three duty days to respond to this action. Any comments or documents you wish to be considered concerning this matter should be included with your response.

█████████████, SSGT, USAF
████████████████

2 – "Memorandum, Letter of Counseling (LOC)" 23 July 2013

DEPARTMENT OF THE AIR FORCE
█ SUPPLY CHAIN OPERATIONS SQUADRON █

MEMORANDUM FOR SRA JAKIA M. ████████   25 August 2013

FROM: SSgt ████████████

SUBJECT: Letter of Counseling

1. On 25 August you came into work with bright red lipstick. When instructed to remove your makeup you stated "its from a slurpee". When asked to provide proof of purchase or evidence you failed to provide adequate documentation.

2. You are hereby counseled! You are in direct violation of Article 92, Failure to Obey an Order or Regulation and lying to a Non-Commissioned Officer. You are to adhere to AFI 36-2903 at all times, If you fail to maintain standards and obey the orders given to you by myself or any others in the chain of command, further administrative action will happen. You have the ability to to perform your duties at and above the standards passed down to you and I look forward to seeing you return to acceptable standards.

3. The following information required by the Privacy Act is provided. **AUTHORITY:** 10 U.S.C. 8013. **PURPOSE:** To obtain any comments or documents you desire to submit (on a voluntary basis) for consideration concerning this action. **ROUTINE USES:** Provides you an opportunity to submit comments or documents for consideration. If provided, the comments and documents you submit become a part of the action. **DISCLOSURE:** Your written acknowledgment of receipt and signature are mandatory. Any other comments or documents you provide are voluntary.

4. You will acknowledge and return this letter to me immediately. You have three duty days to respond to this action. Any comments or documents you wish to be considered concerning this matter should be included with your response.

████████████████████, SSgt, USAF
Supervisor, ████████████, ██SCOS

3 – "Memorandum, Letter of Counseling (LOC)" 25 August 2013

118

DEPARTMENT OF THE AIR FORCE
■ SUPPLY CHAIN OPERATIONS SQUADRON ■

MEMORANDUM FOR SRA JAKIA M. ■■■■   08 October 2013

FROM: SSgt ■■■■■

SUBJECT: Letter of Reprimand

1. On 8 October you took leave for personal matters. On initial feedback you were informed that all leave must be projected at a minimum two months in advance. You indicated that it was for an emergency family situation, but you submitted for ordinary leave. In addition, you left your immediate work area for 45 minutes. When an Non-Commissioned Officer asked you about your absence you stated "I needed a moment, I was in the bathroom crying.: It is to be noted your appearance did not reflect such statement.

2. You are hereby **Reprimanded**! By not following the guidelines set forth in your feedback you have failed to follow my basic standards and expectations and are in violation of Article 92, Failure to Obey an Order or Regulation. I expect you to follow all rules, regulations, and standard in and out of my presence. This shows that you are lacking in your followership abilities and are not adhering to the Air Force Core Values of Integrity, Service Before Self and Excellence in all we do. If you fail to maintain standards and obey the orders given to you by myself or any others in the chain of command, further administrative action will happen. You have the ability to to perform your duties at and above the standards passed down to you and I look forward to seeing you return to acceptable standards.

3. The following information required by the Privacy Act is provided. **AUTHORITY:** 10 U.S.C. 8013. **PURPOSE:** To obtain any comments or documents you desire to submit (on a voluntary basis) for consideration concerning this action. **ROUTINE USES:** Provides you an opportunity to submit comments or documents for consideration. If provided, the comments and documents you submit become a part of the action. **DISCLOSURE:** Your written acknowledgment of receipt and signature are mandatory. Any other comments or documents you provide are voluntary.

4. You will acknowledge and return this letter to me immediately. You have three duty days to respond to this action. Any comments or documents you wish to be considered concerning this matter should be included with your response.

■■■■■■■■, SSgt, USAF
Supervisor, ■■■■■■ SCOS

4 – "Memorandum, Letter of Reprimand (LOR)" 08 October 2013

DEPARTMENT OF THE AIR FORCE
████████████ **AIR FORCE LEGAL OPERATIONS AGENCY**

17 March 2015

MEMORANDUM FOR COLONEL ████████████████/CV

FROM: CAPTAIN ████████████████

SUBJECT: Expedited Transfer – SrA Jakia ████

1. I represent SrA Jakia ████, the victim of an Article 120 offense, as her Special Victims'
Counsel (SVC). I entered into an attorney-client relationship with SrA ████ in February 2015.
As her SVC, I am providing additional information in support of her request for an expedited
transfer.

2. SrA ████'s case was investigated by the Office of Special Investigations (OSI), and the case
proceeded to a court-martial in January 2012. The subject in this case, SSgt ████████████,
was convicted and discharged from the United States Air Force. Since the court-martial, my
client has continuously been subjected to harassment and mistreatment by members in her unit,
including her former leadership[1]. The name-calling and deleterious remarks were so bad that she
filed a complaint with the Inspector General's office in April 2014.

3. The most recent incident occurred this month. In preparation for Sexual Assault Awareness
Month, a link was sent out to all Airmen in which they could search the courts-martial that have
occurred on the ████ AFB installation. Her co-workers again began to make comments about
the outcome of her case. These remarks coupled with the torment she has previously faced, is
what triggered this request for an expedited transfer.

4. I respectfully ask that you approve her expedited transfer. Please afford her the opportunity
to put this experience behind her, and transition to a location where she can receive the support
of her family. I can be reached at ████████, or via email, ████████████. if
you have any questions.

████████████, Capt, USAF
Special Victims' Counsel

Attachment:
Chain of Events, dated 17 March 2015, 5 pages

---

[1] The attached Chain of Events submitted by SrA ████ outlines the various forms of mistreatment she has been
subjected to since the conclusion of the court-martial in 2012.

5 – "Memorandum, Expedited Transfer" 17 March 2015

DEPARTMENT OF THE AIR FORCE
AIR FORCE LEGAL OPERATIONS AGENCY
████████████████

24 March 2015

MEMORANDUM FOR COLONEL ████████████, ████████/CV

FROM: CAPTAIN ████████████████

SUBJECT: Response to Expedited Transfer Denial – SrA Jakia ████

1. I represent SrA Jakia ████, the victim of an Article 120 offense, as her Special Victims' Counsel (SVC). I entered into an attorney-client relationship with SrA ████ in February 2015. On 17 March 2015, SrA ████ submitted a request for an expedited transfer in which I provided a memorandum for record (MFR) in support of the request to her commander, Lt Col ████ ████. On 23 March 2015, Lt Col ████ denied SrA ████'s expedited transfer request. This MFR is to further support SrA ████'s request.

2. Sexual Assault is a broad term that encompasses Article 120, Uniform Code of Military Justice (UCMJ) offenses. In January 2012, SrA ████ testified as the victim of Indecent Exposure, an Article 120 offense. The Subject in her case, a Staff Sergeant, exposed his erect penis to her in the workplace. At the time, she was a 19-year-old, first-term Airman. The Subject was found guilty, and received a bad conduct discharge, two months confinement, and a reduction in grade to E-1.

3. After the court-martial, SrA ████ believed that she would be able to move forward with her life. However, SrA ████, the victim of a sex-related offense, has endured constant mistreatment by her co-workers and chain of command for over three years. Lt Col ████ alleges that SrA Chang's initials being listed on summary material of previous courts is patently false. Yet, with the information contained in the link provided to all members in her unit, readers can easily search for her case, which is exactly what happened[1]. Lt Col ████ references an Inspector General's (IG) complaint filed by SrA ████ in July 2014 as the basis for denying her request for an expedited transfer. However, the IG complaint was filed due to the adverse treatment by her supervisors, TSgt ████████ and SSgt ████████. The request for an expedited transfer is unrelated to the IG complaint; the request was submitted because SrA ████ is a victim of a sex-related offense. Lt Col ████ hasn't spoken with SrA ████ about the court-martial, how it has impacted her, or the expedited transfer request. She was unaware that expedited transfer was an available option until I was detailed as her counsel last month.

4. I respectfully ask that you approve SrA ████'s request. After careful review of the content in her package, you will see that SrA ████ has been re-victimized since 2012. Every

---

[1] Using the information provided in the link, the attached document can be found which summarizes SrA ████'s court case. In the case summary, her initials and the assigned unit are referenced throughout the document.

experience in her unit is a constant reminder of a horrific incident that no 19-year-old serving in our United States Air Force should be subjected to.

5. I can be reached at ████████, or via email, ████████████████ if you have any questions.

6 – "Response to Expedited Transfer Denial" 24 March 2015

DEPARTMENT OF THE AIR FORCE

█████████ AIR FORCE BASE, FLORIDA

22 October 2015

MEMORANDUM FOR A1C JAKIA M. ████████

FROM: ███████ SUPERINTENDENT

SUBJECT: Letter of Reprimand (LOR)

1. Investigation has disclosed that on or about the week of 15 October 2015 at ██████ AFB, Florida, you were insubordinate in the performance of your duties and you willfully provided false information to the 1st Sergeant at the █ Operations Squadron stating you were threatened with AWOL charges by your supervision. In addition, even after being given clear instruction as to how to report your situation by your supervision for accountability purposes you continued to question this units need to be aware of your location and didn't follow the direct orders to provide required paperwork to your Primary Care Manager to validate your situation as required by regulation.

2. **You are hereby reprimanded!** Insubordinate conduct towards non-commissioned officers, providing false statements to other members knowingly and commitment of acts of perjury are all violations of the UCMJ. I hope that you do not fall victim to thinking that the lenient manner in which I am handling your offense gives you a license to continue such behavior or indicates that I condone such behavior. I am, in fact, appalled by your conduct which has discredited you and blemished the image of all airmen. If you do not take this opportunity to prove your potential or if you involve yourself in any further misconduct, I will take much stronger action.

3. The following information required by the Privacy Act is provided for your information. **AUTHORITY:** 10 U.S.C. § 8013. **PURPOSE:** To obtain any comments or documents you desire to submit (on a voluntary basis) for consideration concerning this action. **ROUTINE USES:** Provides you an opportunity to submit comments or documents for consideration. If provided, the comments and documents you submit become a part of the action. **DISCLOSURE:** Your written acknowledgement of receipt and signature are mandatory. Any other comments or documents you provide are voluntary.

4. You will acknowledge receipt of this letter immediately by signing the acknowledgement in the 1st Ind. You have three (3) duty days from the day you received this letter to submit any matters for my consideration. Any comments or documents you wish to be considered concerning this letter must be submitted at that time.

████████, SMSgt, USAF
████████ Flight Superintendent

7 – "Memorandum, Letter of Reprimand (LOR)" 22 October 2015

DEPARTMENT OF THE AIR FORCE
AIR FORCE LEGAL OPERATIONS AGENCY
███ AIR FORCE BASE AREA DEFENSE COUNSEL

23 October 2015

MEMORANDUM FOR ALL REVIEIWING AUTHORITIES

FROM: ███/ADC (Capt ███████████)

SUBJECT: Letter of Reprimand, SrA Jakia M. ███

1. I am the detailed defense counsel for SrA Jakia ███ and I am writing this letter on her behalf to request that the Letter of Reprimand (LOR), dated 22 October 2015, be rescinded.

2. The LOR indicates that SrA ███ is being disciplined for four things: 1) being "insubordinate in the performance of [her] duties;" 2) "willfully provid[ing] false information to the 1st Sergeant at the ███ Operations Squadron [by] stating [she was] threatened with AWOL charges by [her] supervision;" 3) questioning the "units (sic) need to be aware of [her] location;" and 4) not "follow[ing] direct orders to provide required paperwork to [her] Primary Care Manager to validate [her] situation."

3. As it relates to the first allegation, the LOR is not specific about what exactly SrA ███ did that was "insubordinate in the performance of [her] duties." As a rehabilitative tool, the LOR should specifically state what SrA ███ did that was insubordinate so that she can correct that behavior in the future. Failing to state what she did wrong also limits her ability to effectively respond to the LOR. However, based on the LOR as a whole, I assume what is being alluded to in the first allegation is that SrA ███ was insubordinate when she did not allow members of the unit to visit her in the hospital and would not provide detailed information about her medical condition. Since she was hospitalized for five days, it is unclear how any of this conduct could fairly be characterized as "in the performance of [her] duties." But more importantly, SrA ███'s conduct was wholly within her medical privacy rights. DoD 6025.18-R covers what medical information a military medical treatment facility (MTF) may release. In deciding what to release, MTFs must follow the minimum necessary rule (DoD 6025.18-R, C8.2.1) which requires that they release only that information that is minimally required for the unit to know. It's worth noting that this is a requirement for military MTFs, not civilian MTFs or the patient. Civilian MTFs follow civilian privacy laws. SrA ███ was only required to let her unit know of her status (i.e. that she was hospitalized), which she clearly did based upon the text messages. There is no requirement for her to share all the intimate details about her medical condition or allow people from the unit to visit her while she is in the hospital. If the unit wished to know more information about her condition then it should have followed the correct protocols and requested that information through the MTF. In sum, SrA ███ cannot be disciplined for insubordination for acting in accordance with her medical privacy rights.

4. The second allegation, that SrA ███ provided false information, is equally without merit. As she notes in her response, SrA ███ never even communicated, directly or indirectly, with the first sergeant of the ███ Operations Group. This fact can be confirmed simply by talking to the Shirt. What appears to have happened is SrA ███'s husband talked to the Shirt (who happens to be his first sergeant and not SrA ███'s) about what was going on with his wife. In this conversation he may have said something that would have the connotation of what is alleged

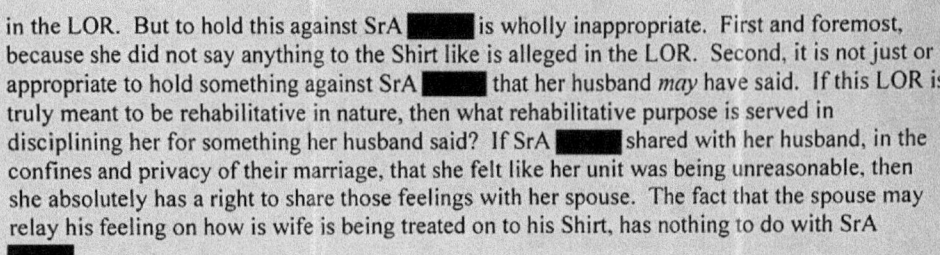

in the LOR. But to hold this against SrA ▮▮▮▮ is wholly inappropriate. First and foremost, because she did not say anything to the Shirt like is alleged in the LOR. Second, it is not just or appropriate to hold something against SrA ▮▮▮▮ that her husband *may* have said. If this LOR is truly meant to be rehabilitative in nature, then what rehabilitative purpose is served in disciplining her for something her husband said? If SrA ▮▮▮▮ shared with her husband, in the confines and privacy of their marriage, that she felt like her unit was being unreasonable, then she absolutely has a right to share those feelings with her spouse. The fact that the spouse may relay his feeling on how is wife is being treated on to his Shirt, has nothing to do with SrA ▮▮▮▮.

5. The third allegation is closely related to the first allegation because it centers on SrA ▮▮▮▮'s reluctance to provide specific details about her medical condition and to allow non-family to physically visit with her- both of which she is legally entitled to do. SrA ▮▮▮▮ provided details about her hospitalization to several members within the unit. MSgt ▮▮▮▮ even went to the hospital where they confirmed she was there. To try to further satisfy her unit, SrA ▮▮▮▮ had the hospital draft up a formal note stating she was in the hospital. It's unclear what more SrA ▮▮▮▮ was expected to do without wholly waiving her medical privacy rights.

6. The fourth and final allegation is simply incorrect. As noted in her response, SrA ▮▮▮▮ tried to set up an appointment with her primary care manager (PCM) while she was still in the hospital, despite the fact that she is not required to do so until after she is discharged (and she was informed of this fact when she tried to make the appointment). SrA ▮▮▮▮ met with her PCM at the first available appointment and provided all of the required documents.

7. When looking at how SrA ▮▮▮▮ behaved during the timeframe in question it is important to look at what was going on with her medically. She was pregnant, vomiting constantly, and being held at the hospital. Anyone who has ever been at a hospital knows that it is a stressful and scary experience. She was barely clothed and only wearing a gown. She was hooked up to monitors and IVs. Having blood drawn and tests performed. And to make it all worse, she contracted an illness while in the hospital. All of this persisted over a period of five days. It is understandable that she would not want to see people other than family. It is understandable she did not want to share intimate details about what was going on with her health at that moment. What is not understandable is why she was expected to give into every one of her unit's unlawful demands as she sat bound to a hospital bed.

8. From the allegations in this LOR it is clear that the LOR is personal, and not truly meant to rehabilitate SrA ▮▮▮▮ for any disciplinary issue. It is an attempt to punish her for not capitulating and doing exactly what the unit wanted her to do- despite the fact that she had no legal obligation to do so. For the reasons above, I request that the LOR be rescinded and this issue resolved at this level so it does not necessitate being resolved at a higher level.

▮▮▮▮▮▮▮▮▮▮▮▮, Capt, USAF
Area Defense Counsel

8 – "Memorandum, Letter of Reprimand (LOR)" 23 October 2015 (2 of 2 pages)

124

DEPARTMENT OF THE AIR FORCE
██ AIR MOBILITY WING ██
██ AIR FORCE BASE, FLORIDA

27 Oct 2015

MEMORANDUM FOR RECORD

FROM: ███████████ █ ███

SUBJECT: Jakia ████ Recent Medical History

1. The purpose of this letter is to document the recent medical history of Jakia ████ in an effort to clarify her condition and treatments.

2. Jakia ████ was diagnosed with pregnancy on 22 September 2015. She was reporting nausea and some vomiting at that time, as well as abdominal pain. She was seen again in clinic on 29 September 2015 and sent to the emergency room (ER) to rule out an ectopic pregnancy. She was cleared for this concern, but remained nauseated and continued vomiting until her next appointment on 2 October 2015. She was dehydrated at that time. We were unable to start an IV for adequate rehydration and sent her via ambulance to the ER. She was kept there for rehydration and then sent home to follow up with her Obstretician (OB). She then had an appointment with her OB on 8 October 2015. The patient then returned to the ER with intractable nausea and vomiting on 12 October 2015 and remained admitted until 16 October 2015. She followed up with her OB as instructed on 20 October 2015, who recommended she return to the hospital for more treatment or stay home all week to recover. She was put on quarters by our office to facilitate the OB recommendation until we could gather more information from the OB office. The patient brought in the OB note on 22 October 2015, and we again recommended patient return to the hospital as she was very dehydrated by the time we assessed her. She did proceed to the ER for further treatment. We have documentation for all these visits and treatments in the patient's medical record.

3. Should you have any questions, please contact me at ███████████████████████.

███████████████ USAF, ███
Physician Assistant, Family Health
Family Health Clinic, ██████ AFB

Image 10, Next Page: An excerpt from my Inspector General complaint against my previous chain of command. During this time, I was hazed, bullied, and retaliated against for making a protective communication regarding the maltreatment my unit was subjecting me to. Although my direct supervisor gave me the highest rating possible on my annual report, my Commander and Flight Chief non-concurred. In doing so, they wrote derogatory remarks which had the potential to affect my career. Just to name a few, it is noted that during this time I received high recognition for saving a life, won Distinguished Graduate from Airman Leadership School (top 10% of my class), deployed, awarded an Achievement Medal, and received my Community of the Air Force College degree. In addition to the remarks, my First Sergeant electronically sent my Personnel Information File to my gaining unit in a malicious attempt to attack my reputation. With the attached findings and multiple witness statements, my case was returned unsubstantiated.

d. Maj ██████ and MSgt ██████ nonconcurred with the Complainant's rating on her
    January 31, 2019, EPR. MSgt ██████ and Maj ██████ signed that EPR on March 1,
    2019. Their comments explaining their nonconcurrence with the Complainant's
    EPR qualify as unfavorable personnel actions as defined in DoDD 7050.06 because
    they affected or had the potential to affect the Complainant's career. Maj ██████'
    and MSgt ██████'s comments raise doubts about the Complainant's character,
    readiness for promotion, and her ability to succeed at the next higher rank.
    Additionally, MSgt ██████'s decision to transmit the Complainant's PIF to her new
    unit was an unfavorable personnel action because the information in the
    Complainant's PIF revealed to her new supervisors and commander the errors and
    potential misconduct the Complainant committed, and therefore had the potential to
    affect her career.

████████████████

████████████   7

    i. MSgt ██████, as the additional rater on the Complainant's EPR, nonconcurred
        with the Complainant's EPR and included the following comments:
        "Disagree with the rater's assessments. While [the Complainant] progresses
        in aspects of her duty performance, she requires growth as an NCO in
        professional conduct, leadership of airmen, effective communication, and
        decision discernment." MSgt ██████'s nonconcurrence and comments qualify
        as a personnel action because they were incongruent with the Complainant's
        evaluation from ██████, and were likely to cause confusion and raise
        doubts about the Complainant's readiness for promotion, and therefore had
        the potential to affect her career.

    ii. Maj ██████, as the reviewer on the Complainant's EPR, also nonconcurred
        with the Complainant's EPR and included the following comments: "I concur
        with additional rater comments, in addition to substantial improvement is
        required in Air Force core values." Like MSgt ██████'s nonconcurrence and
        comments, Maj ██████' nonconcurrence and comments were likely to cause
        confusion and raise doubts about the Complainant's character and ability to
        succeed at the next higher rank, and therefore had the potential to affect her
        career.

    iii. MSgt ██████'s decision to transmit the Complainant's PIF to her new
        command revealed to her new supervisors and commander the errors and
        potential misconduct the Complainant committed. It is possible her new
        supervisors and commander could consider that information in decisions
        they make with respect to the Complainant's assignments and whether or
        not to support her future promotions, and therefore this decision had the
        potential to affect the Complainant's career.

10 – "Inspector General Complaint"

DEPARTMENT OF THE AIR FORCE

████████████████

19 May 2020

MEMORANDUM FOR ████ /CC
SSgt Jakia Lindley

FROM: ████ /CC

SUBJECT: Central Registry Board (CRB) Incident Determination

1. The CRB met on 19 May 2020 to review incident ████████ involving SSgt Jakia Lindley. The allegation was adult emotional maltreatment, sexual maltreatment of SSgt Jakia Lindley by TSgt ████████. The board determined the incident **met** the criteria for sexual maltreatment, **did not meet** the criteria for emotional maltreatment. Only met criteria incidents will be reported to the DoD Central Registry database.

2. IAW AFI 40-301, if the alleged offender or victim disagrees with the determination, either may request an Incident Status Determination Review (ISDR). The request must be submitted in writing to the Family Advocacy Office within 30 days of notification of the CRB determination. The Family Advocacy Office will present the ISDR request to the CRB Chairperson. An ISDR may be granted when there is new information that was not presented to the CRB, and that information could affect the outcome determination. An ISDR may also be granted if the committee failed to comply with the CRB's published directives and standards. A signed and dated copy of this letter must be attached to the request.

3. The CRB findings are to be shared between you and SSgt Lindley. This letter serves as verification of the date of notification. Both of you must sign and date this letter and provide a copy to SSgt Lindley as a record of the CRB determination date. Questions may be addressed to the Family Advocacy Office at ████████████████████.

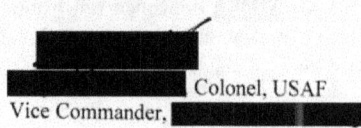

████████████████ Colonel, USAF
Vice Commander, ████████████

11 – "Memorandum, Central Registry Board (CRB) Incident Determination" 19 May 2020

DEPARTMENT OF THE AIR FORCE
██ **MEDICAL OPERATIONS SQUADRON** ██

MEMORANDUM FOR RECORD.

FROM: Family Advocacy Program

████████████████

SUBJECT: Expedited Transfer of Military Service Member

1. SSG Lindley has been receiving Family Advocacy Program treatment services following an incident of domestic violence and sexual assault with her ex-husband. SSG Lindley has an open unrestricted maltreatment case with the Family Advocacy Program.

2. Currently, SSG Lindley does not have a support system in place due to the geographical separation with her current husband, ████████, who is also active duty in another state. SSG Lindley's mental health and emotional stability will benefit from have a support system in place, while going through the trial for her sexual assault.

Family Advocacy Officer

Image 13, Next Page: From my own medical records, when SJ allowed us to attend marriage counseling. My therapist noted my anxiety and fear when he is angered, and provided me with techniques on how to de-escalate the situation when he is triggered (as SJ had a history of controlling and aggressive behavior). Although SJ self-reported aggressive borderline behaviors, I believed in him that things would get better.

PROGRESS NOTE ████████ Counseling, LLC

Client Name: _Qabia_____ Diagnosis _████████_

Progress Note: □ Significant change in med. condition and/or medications. □ Slight Change

□ No Change □ Significant change in mental status. ☑ New stressors and/or extraordinary events.

Description: _Former Clients of this therapist seen Under Dual Counseling._

Target Problem or issue of session: _Spouses & In Anxiety & aggression. Control_ _Issues as described by Ct. & Spouse being "ongoing."_

Revised or New Goal: _Provide De-Escalation techniques as means to lower anxiety_ _driven altercations = spouse. + ████████_

████████████████████████████

Risk Assessment: □ None □ Suicide □ Homicide □ Phys. or Sexual ☑ Domestic Violence

Explanation of above if risk is present: _aggressive, borderline behaviors Self-_ _reported by Spouse. States he is willing to work towards goal_ _████ of harmony in the marriage_ Date & Time _4/5/__

████████████████████████████

████████████████████████████

Revised or New Goal: _Provide Spouse c̄ Information re: her Anxiety & fear when he is (Ct.) angered._

Specific Strategies, Interventions, and update: _Discussed best_ _Strategies for de-escalating problematic comm. Patterns_ _c̄ (Op) family of origin + c̄ Spouse when he is escalated_

Risk Assessment: □ None □ Suicide □ Homicide □ Phys. or Sexual ☑ Domestic Violence _due to fin. Stress_

Explanation of above if risk is present: _Hx of controlling_ _c̄ family_ _+ aggressive behaviors._ _Business_

Signature of Clinician _████████_ Date & Time _Phone Call._ _4/13/__

13 – "Medical Records, Progress Note 1"

Image 14, Next Page: As things progressed in our marriage, it became apparent that things were escalating. My therapist has now started to provide "safety" plans for me. During these sessions it was imperative to find the triggers that caused these "aggressive tactics" by SJ that cause me to "flee residence."

PROGRESS NOTE ███████ Counseling, LLC

Client Name: ___O___abia_____ Diagnosis ___███████___

Progress Note: ☐ Significant change in med. condition and/or medications. ☑ Slight Change

☐ No Change ☐ Significant change in mental status. ☑ New stressors and/or extraordinary events.

Description: ███

Target Problem or issue of session: ████████████████████████████

Revised or New Goal: ☓ In anxiety. Identify triggers in Spouses Interactions that lead to Nabia fleeing residence.

Specific Strategies, Interventions, and Update: ███

Risk Assessment: ☐ None ☐ Suicide ☐ Homicide ☒ Phys. or Sexual ☒ Domestic Violence

Explanation of above if risk is present: Spouse's anger & inability to provide Ct. the time & space she needs, to calm herself.

Signature of Clinician ___███████___ Date & Time___4/26/13___

New Progress Note: ☐ Significant change in med. condition and/or medications. ☐ Slight Change

☐ No Change ☐ Significant change in mental status ☑ New stressors and/or extraordinary events

Description: Processed rape & assault at age 15 by Childhood friend.

Target Problem or issue of session: Identifying behaviors that are experienced by Ct. from Spouse to be aggressive and/or, Controlling so that Safety can be established.

Revised or New Goal: ███

Specific Strategies, Interventions, and update: ████████████████████████ Specific Examples of What Constitutes abuse ie verbal emotional, physical.

Risk Assessment: ☐ None ☐ Suicide ☐ Homicide ☐ Phys. or Sexual ☒ Domestic Violence

Explanation of above if risk is present: aggressive tactics by Spouse ie. Chasing upstairs, picking lock in closet, etc. that escalate altercations that are not considered ███████ by Ct. to be major Issues.

Signature of Clinician ___███████___ Date & Time 5/10/13

14 – "Medical Records, Progress Note 2"

PROGRESS NOTE ▮▮▮▮▮▮ Counseling, LLC

Client Name: _____ Jakia ▮▮▮▮▮ Diagnosis ▮▮▮▮▮▮▮

Progress Note: □ Significant change in med. condition and/or medications. □ Slight Change

□ No Change □ Significant change in mental status. □ New stressors and/or extraordinary events.

▮▮▮▮▮▮▮▮▮▮▮▮▮▮▮▮▮▮▮▮▮▮▮▮▮▮▮▮▮▮▮▮▮▮▮▮

Revised or New Goal:

Specific Strategies, Interventions, and update: Continue Check Ins. Continue leaving Residence When Ct. Senses that Spouse is escalating.

Risk Assessment: □ None □ Suicide □ Homicide □ Phys. or Sexual ☑ Domestic Violence

15 — "Medical Records, Progress Note 3"

Image 15, Above: At this point, the only suggestion was to continue check-ins and continue to leave my residence whenever I sense SJ is escalating.

Chapter 18

Please Let Me Die

"I've tried so hard to use those two annoying F words; forgive and forget. It never works."

~ Calia Read

AS IF THINGS couldn't get any worse, I threw my back out and had to see the doctor. After examinations and X-rays, the doctor sent me to physical therapy. On my first visit, the therapist moved my leg in a certain position, and I screamed like he'd set my body on fire.

"Did you get an MRI?" he asked.

"No," I told him. "I got an X-ray."

"Okay," he shook his head. "We're done here. You shouldn't be doing physical therapy, especially with a back injury, without getting MRI results first. I might do more harm than good. I'm sending you back to your doctor to schedule an MRI."

The MRI showed what the X-ray couldn't: a bulging disk that had herniated. My right butt cheek felt like it belonged to someone else that didn't want to move when I'd try to walk. The pain radiated down my leg. The doctor sent me home with no treatment, guidance, or referral to a specialist.

Later, at home, I was changing my clothes when my back locked up. My top half fell across the bed, and my legs buckled underneath me as my knees reached the floor. I looked like I had fallen asleep saying prayers before bed. The pain was so intense, it took me several minutes to scream out to SJ for help. Once he saw me bent over in pain, he called 911. Once the paramedics arrived, they strapped me to a sitting stretcher. I passed out before they got me in the ambulance. At the ER, they administered morphine, and the pain subsided.

This back problem kept me out of work for a month or so. Once I got stronger, I got back into physical therapy. But I still couldn't walk without wobbling like a weeble. No position remained comfortable for long. I found myself shifting from side to side when the pins and needle sensation returned. I tried to keep my suffering to myself, not wanting to complain to SJ, who already had his hands full helping me shower, get dressed, and walk to and from the bedroom, bathroom, or couch whenever I needed to move.

One morning after helping me out of the shower, SJ decided the time had come for me to pay him back for all his help.

"It's been a long time since we've had sex," SJ said as I stood in the closet, picking out my clothes for the day. "Too long. I think we need to end this dry spell."

"SJ," I whined, hoping he was making a sick joke. "My legs are numb. I'm in pain. I can't have sex."

He shook his head, instantly enraged. "So you're telling me no? Are you fucking kidding me? I've been helpful, and I've been patient. But enough is enough," he said. His tone had turned like a radio station being flipped to a different music genre. There was no telling if—or when—it could be turned back. And this channel only played dark and chaotic music.

One moment I'm standing before my clothes, and before I could absorb what was happening, he shoved me to the ground and sat his weight on top of me, knocking the wind out of me. I couldn't breathe, and his weight put pressure on my herniated disk, radiating pain throughout my torso. I started to black out but knew I had to hold on.

"Is this what you want me to do?" he yelled, simultaneously slapping himself in the face. "Is this what you like? Does this make you happy?"

"Stop," I panted. "I can't breathe." Tears covered my face as the pain became overbearing. His weight compressed my spine, overwhelming my ability to think or function.

"The Bible says, 'The wife hath not power of her own body, but the husband!' That means you can't deny me sex! It's in the Bible!" he screamed.

I continued to cry and grunt in pain, trying in vain to twist and writhe in any direction to relieve the pain—or escape him. "Stop it. You're hurting me," I sobbed, feeling more defeated with each breath but still hoping I could reach something inside of him.

He continued quoting the Bible over and over, but not like any minister I'd heard. "The wife hath not power of her own body, but the husband," he repeated, using Scripture like a weapon to terrorize me. "Your body belongs to me! It's mine, for my pleasure! By God's hand, you must submit to me."

"Please," I begged him again, still hoping to appeal to any goodness in him.

"Will you have sex with me?" he asked again.

"No," I said.

"Will you have sex with me?" he repeated, getting louder.

"No," I answered.

He repeated the same verse, continued making the same request, and kept slapping his face, then slapping the floor next to my head, narrowly missing my face. The message was clear: he wanted to hit me, and it might not be long before he did, if I didn't give in.

Finally, he broke me.

"Fine," I answered. I closed my eyes, resigned that I couldn't make him stop— nor could I escape. But it was *not* fine; *I* was not fine. I had no choice in what he'd do next.

He ripped open my robe, exposing my body fully. His eyes looked cold as he stripped off his clothes. He then raped me on the floor as my soul felt like it left my body. Except for the pain. The pain kept me conscious of what he was doing. Words of scripture that he hurled at me to justify his actions echoed inside my head. I tried to take my spirit somewhere else far away, but the dagger in my back kept bringing me back to the present.

Once he finished an act that only involved him using me for what he needed, he got up. Then he picked me up like an object he was finished using, carried me to the bed, and dropped me there like a heap of dirty laundry for someone else to wash. But I wasn't washable. I felt permanently stained.

He left the room without saying a word, leaving me in spasms of pain, brokenness, and utter defeat.

He went to the kitchen to make himself something to eat. I wouldn't have been surprised to hear him humming as he made a sandwich; the event seemed so inconsequential to him.

Meanwhile, I lay on the bed with my heart on fire. I felt like I was bleeding out all hope, like I was nearly empty. The moment that I stopped saying no was the moment that I lost myself. And I couldn't live like this. *I wanted to die.* My life had become an endless chapter of pain, shame, humiliation, and abuse.

Recently, I found a journal I kept hidden from that period of time. Here's what I wrote:

"I was so excited for his return, now I just wish he'd go. This cannot be love. He wields this Bible in front of him, but he is not a holy man. Even the devil was once an angel. Today I hate my skin. I scrubbed away at my skin so much it burns. This cannot be love. This cannot be God's will. If it is, I do not want to draw closer to this being.

"He says you cannot rape your wife, but why do I feel like this? Why do I feel like a stranger was just inside me? Whenever I say no, he uses God's words to command me. This God you quote is a sick man. If he loves us, then God, why am I being punished? What did I do? Why would you let a stranger into my home? Is this truly the man I married, or is this a test I must go through? Again and again I watch over my body from above while he takes what he wants. What am I to learn from this?

"If this is what my life has become, I pray you remove me from this hell. I no longer want to feel this pain anymore. He won't allow me to leave, so I pray for you to come take me, please Lord."

Shortly after raping me, SJ flipped the dead bolt on our front door, so it locked and unlocked from the outside. While I convalesced at home, he'd lock me inside when he left for work. Then he unplugged the internet cable and took it with him each day. I couldn't walk, leave the house, call anyone, or go online. I could scream, but no one would hear me. I could do nothing except wait for him to get home. Since I struggled to walk unassisted, SJ left me on the couch with snacks and just enough water to keep me alive—but not so much that I would need to get up to pee until he returned home.

If I questioned him or dared to fight back, he hurt me more. I retreated into a tiny compartment within my soul, a safe place, one with a switch that allowed me to shut off my mind and feelings. I learned to flip that switch at will, giving me escape from him. In the safety of my little mental construct, I watched SJ rape my will, freedom, choices, and body. Floating above the tragic scene below, I detached as if I were watching a television show.

At first, I kept count of how many times he raped me. I wrote down the details in my hidden journal, hoping that process would reveal an answer. But then I figured: *What's the point? What can I do about it?* Then I stopped counting and journaling about it. Since I was in hell, I didn't need reminders. It wasn't like I could forget anyway.

Around this point in our marriage, SJ, seeing how miserable and despondent I'd become, agreed to couple's counseling. We started joint therapy through a military counseling service. Not only did the therapist help me feel less alone with my troubles, but she also pulled no punches with SJ.

"What you're doing to her, SJ," she said, staring him straight in the eyes, "is not okay. If my daughter were in a relationship with someone like you, I would be terribly upset, and I would want her to leave for her own safety," she told him while I sat there feeling heard. Then she added, "I don't normally encourage divorce, but if you two can't resolve this, I think you need some time apart from each other. Because one of you will get hurt if you keep on this path."

SJ had little to say on the way home, except that he found counseling a waste of time and didn't want to attend more sessions. But he ended up coming with me a few more times. When the counselor recommended we have individual sessions, he put his foot down.

"Nope," he said, making sure I understood there would be no further discussion. "You're not going. I'm not going. We're done."

He feared what I might say if I were alone with the therapist.

During our first meeting, she told him that he was a textbook case of narcissism with traits of a sociopath. I was not surprised. She checked off domestic violence, sexual abuse, emotional abuse, and physical abuse. When I first saw what she wrote, I thought: *My God! Why didn't she call the police? I could have been killed!*

For a time, she gave us homework, like *write nice notes to one another* and *say nice things to each other every day*. I did my homework. SJ did not. He found all that kindness a waste of time. On top of that, I'm sure that he told himself that I was the one with a problem, so why should he be inconvenienced?

Once we stopped seeing a counselor, our relationship returned to normal, meaning it went back to hell. The old routine returned.

SJ rarely made overtures to help me stay with him. He did, though, put a lot of energy in cutting off my escape routes. He reminded me that I didn't have any money, since he controlled all the bank accounts. I didn't have any close friends except Vickie, and my drama with SJ wore her out. Ironically, Vickie grew frustrated with my willingness to stay in the marriage when I'd only told her the G-rated parts of how SJ controlled me. Had I told her he physically and sexually abused me, I'm not sure if she would have tried to kill him or slapped me into action. I had no one to turn to, no one to even talk to. He reminded me that I had a shit reputation at work for being a rat, so no one would believe me if I tried to turn in my own husband. It would come down to his word against mine.

Trapped, I stayed in my marriage, with a man that didn't respect or even like me. He was like a man that claimed to love animals, so he kept stuffed heads on his wall. He played the great hunter, and I, his reluctant trophy, stayed locked inside the house—on display just for him. He became the strict, controlling father while he treated me like the rebellious, broken daughter.

Meanwhile, my supervisor at work let me know that unless I died, I needed to be at work. She wrote: "You need to take charge of your health. If it's not an emergency, you need to get back to work. Your coworkers consider you a weak link because you cannot keep up, and they must do your work. If you can, come back and get on the right track."

Of course, she wouldn't know that I'd been locked up, unable to communicate with the outside world, and occasionally raped. But on that point, SJ never considered what he did *rape*. Whenever I tried talking with him about it, he told me that you can't rape your own wife, you can only "take advantage" of her. He said the Bible was clear in that regard. He was the man, and my body belonged to him. He was within his rights, by God, to do whatever he wanted to me.

I wasn't scamming the system and playing sick. I couldn't walk for a time because of the pain in my back. The abuse at home got so bad, I would have crawled into hell had that been an option. *Oh, wait. I can't. Because I'm locked inside my damn apartment.* So instead of arguing or trying to defend myself at work, I begged my neurologist to clear me to go back to the office, which he did under strict limitations and increased PT visits.

In hindsight, I should have followed his orders. The first day I returned to work, I fell within two hours. I ended up back in the ER, and this time I got put on extended quarters, which is military speak for sick leave. As a *thank you* for trying to come back to work and getting hurt again, I got paperwork for being a work hazard.

Shortly after returning to work, my common access card—the magnetic stripe card that gives you access to any areas within your clearance level on base and the DoD computer network—stopped working. This occurs so often they have a procedure to follow when it happens: finish your current shift helping others with their work, then get your card restored the next day before returning to duty. Except that's not what my supervisor told me to do when I reported my card deactivated.

"I want you outside picking weeds from the parking lot," she told me.

"Ma'am, it's over 90 degrees outside," I reminded her, since we had been outside in the early morning pulling weeds when it was a cool 75 degrees.

"That shouldn't be a problem for you," she waved off my concern. "Your people did it back in the day."

"I'm sorry," I tensed up. "Would you repeat that?"

"Oh, my goodness, Airman Clark," she said. "Why are you always so angry? You just need to calm down."

"Ma'am," I said while holding back my emotions, "you just told me to pick weeds in the blazing heat since my slave ancestors did it."

"Oh my God," she sighed. "No, I didn't. What are you even talking about?"

Fuming with anger and not wanting to make the situation worse, I went outside before I acted on my emotions. I worked in the heat, bending down repeatedly and straining my back even more. But I didn't feel physical pain. The heat I felt came from inside of me. As soon as I finished, I went inside and filed a report with the inspector general for my supervisor's racially inappropriate comment.

My supervisor had no problem creating a plausible scenario for the investigators. "Dear God, no," she told them. "I would never say something like that. What happened is that we were all outside earlier in the morning picking weeds as part of base beautification. That's all that happened. Then I sent her outside in the afternoon to finish up. I planned to join her and bring other airmen. But last-minute, urgent work came in, and I lost track of time."

Which was bullshit. I had been outside early in the morning with everyone else. But six hours later, in the heat of the day, she sent me out there again—and alone, since "my people" were accustomed to those kinds of working conditions.

Instead of a reprimand, they recommended that in the future, the next time she sent airmen out in the heat, she made sure they had proper gloves and water, especially in extreme heat.

I needed to get out of this place for a while, so I put in a request for leave, which my supervisor granted. SJ and I took a short trip to New York for downtime. While checking in for our return flight home, another plane had to make an emergency landing. They closed my runway, then cancelled all flights in and out of the airport.

"Ma'am," I reported in from New York. "I'm not sure if you caught the news, but they've cancelled all flights out of this airport for the time being. I just wanted to let you know that I'm delayed returning to work."

"Not my problem, airman," my supervisor replied. "Rent a car and drive back so you're here in the morning."

"I'm in New York," I explained. "If I left now and rented a car, I still couldn't get to Virginia by morning. They've rebooked me on the earliest projected flight that will get me back to base."

When I returned, she gave me a letter of counseling for "taking a last-minute leave," even though I gave her two weeks' notice. And she wrote me up for not being back to work on time when the airport shut down.

In the military, there's a chain of command available to use to resolve conflict. After so many clashes with my supervisor, I requested a meeting with her boss. He listened, nodded, asked good questions, and seemed to understand that how my supervisor treated me was not in line with their policies. As a result, he told me that he would talk to her.

Of course, I couldn't sit in on the meeting he held with her, but I knew immediately after the two of them had met.

"If anyone in this section goes to my boss again without coming through me first, you will receive automatic paperwork," she announced to our entire section. "Do I make myself clear?"

All the airmen she had a close relationship with immediately spoke up, saying things like, "We would never do that!"

Then she looked directly at me. "Airman Clark, would you have a problem with that?"

"No, ma'am," I said like a chastised child.

"Good," she concluded. "Since everyone else in the office agrees not to go over my head again, then you should be in agreement, too."

"Yes, ma'am," I replied.

Cutting off an airman's access to go above a supervisor's head is illegal. You can't punish an airman for using the chain of command. We're supposed to have open communication, the old "see something, say something" policies to avoid a scandal like Tailhook in the early 1990s. If you don't remember Tailhook, several United States Navy and US Marine Corps aviation officers at a convention in Las Vegas allegedly sexually assaulted more than eighty women and seven men. Had any junior officer present felt comfortable speaking up, this situation could have been prevented.

When my leader singled me out in front of the office, she put a target on my head for more harassment. She might as well have made me wear a T-shirt with a picture of a rat on it.

First sergeants serve as informal "therapists" for airmen, since they have tenure and deep understanding of policies, procedures, and politics within the ranks. So instead of going over my supervisor's head, I used my first sergeant as an unbiased sounding board.

"My supervisor keeps bringing up the court-martial of Morin," I told him. "And she's coming after me. She's made it so I can't report her no matter what she does. All I want is a fresh start."

"In my opinion," he told me, "I think you need thicker skin if you're being affected by the people around you and what they say and do to you. You reported S.Sgt. Morin for what he did. If you can't deal with the backlash from that, then you shouldn't have reported him."

I made my bed, and now I need to lie in it, was how I interpreted what he said.

Soon afterwards, my supervisor added more disciplinary action to my record for using my work email at work to check my work tasks. Our office had two sections: computer ops and personal desk side. When you're on the op side, once all work is done, you can check your email. Once I did all my work for the day, I checked my email on the ops side to look at my work tasks. The moment I pulled up my email, she wrote me up on the spot. Then the next day, I got paperwork again for *not checking* and meeting the requests sitting in my inbox.

I'm damned if I do, damned if I don't. I can't win here! I got paperwork for trying to check my email, and then I got paperwork for not checking my email.

Around that same time, my grandfather fell ill. He had been in declining health with a terminal diagnosis, and Mom let me know that he was in his final days. I requested time off to see him.

"Request denied," my supervisor said without hesitation. "He's not my grandfather. And you are aware that the office policy is two months' notice for leave. Refer back to the feedback that you signed."

He died soon afterwards. I never got to see him before he passed.

A colleague saw me crying at my desk when I got word of his death.

"Go to the funeral," he said. "I can approve your time off. You need to be with your family. If there's any heat, I'll take it."

This colleague was one of three people who could grant me leave for personal reasons, and I was so grateful that he stepped in so I didn't have to talk to my immediate supervisor.

When I got back, I received paperwork for taking last-minute leave.

"You should have better anticipated his death so you could have made your request for leave within the two months' policy," my supervisor told me. "Your last-minute leave impacted the mission."

Work was hell. Home was hell.

I needed a change. *But what?*

Chapter 19

My Turn

"Things to do today: 1. Get up 2. Survive 3. Go back to bed."

~ Unknown

IN EARLY 2014, I learned that my recruiter's comment, "They don't like to deploy females," was about as accurate as many of the other things he'd told me. I got orders to deploy to Afghanistan. But this time, I wasn't mad at him for bending the truth. *Send me into battle! Put me on the front line! I don't care anymore. Just get me the hell out of Langley AFB, away from my supervisor, and away from SJ,* I told myself.

"Cancel it," SJ said when I told him my news.

"You know," I shook my head, "you don't just cancel deployment. When they call you up, you go."

"I'm going to find a way to get you out of it," he said. "Deployments are bad. That's when people—lots of things happen on deployment, and I won't have my wife in that environment. I've been on deployments, and I know how they are."

I knew that what they said about deployment is what they say about Vegas. What happens on deployment stays on deployment.

"They are going to try to get you to divorce me," SJ continued. "And they're going to try to sleep with you."

Very telling that *his first thoughts about me deploying were not about my safety but about me sleeping around. Good to know. I feel so loved.*

SJ knew he couldn't keep me from deployment, so he made me enter a contract of sorts. He read aloud from Genesis 31 where Jacob separates from his father-in-law, Laban, taking his wives, Leah and Rachel, with him. Laban and Jacob agree to build a pile of stones, a figurative watchtower, to commemorate a promise between the two men. Laban would agree to let Jacob leave with Leah and Rachel, but in return, Jacob had to promise not to abuse the two women or take additional wives.

In Hebrew, this watchtower is called a *mizpah*. Since no other individuals were present to witness this covenant, they entered into their agreement with God as their witness.

SJ wanted us to get a tattoo of our holy covenant, but I felt like a tattoo was too permanent for such a temporary deployment. Finally, SJ compromised and got us both bracelets with a Scripture on it: Genesis 31:49, "May the Lord keep watch between you and me when we are away from each other."

Call me old fashioned, but I didn't need a tattoo or bracelet to remind me not to commit adultery. The way I looked at SJ's request is that he wished to brand me like a side of beef.

Then he insisted I get baptized, too, so I would remain free from sin. He explained it this way: "If you sin after you get baptized, you go straight to hell." He was telling me as a not-so-subtle reminder to keep my knees together. And the love embedded in his rationale was: "Get baptized, not to go to heaven if you die, but so you'll go straight to hell if you sin!"

I deployed in April of 2014, spending my twenty-second birthday in Shannon, Ireland. I sang "Happy Birthday" to myself to celebrate not just the day of my birth, but the day of my freedom from SJ and my toxic work environment.

When we finally touched down in Afghanistan, its beauty surprised me. Here was a country under constant upheaval with civil wars, the Taliban, al-Qaeda, and continual bombing raids, but it looked like paradise to me. While I stood outside looking at the beautiful mountains, I heard a siren followed by explosions.

"Okay," I said as I ran for cover. "Those were fucking bombs!" For the first time, I understood why my family had concerns about me deploying there.

Bombs exploded around us every day I was in Afghanistan, but I never got used to them. I felt scared whenever the ground shook. But then I'd laugh and think to myself, *SJ's worried about me hooking up, and all I can think about is keeping my ass attached to my body.*

Besides being afraid for my life, I found several things to love about deployment. For the first time since childhood, I had responsibility for myself alone. I didn't have to babysit a drunk boyfriend, manage the emotions of my man-child husband, clean house, or play hide-and-go-hide-some-more from evil coworkers and supervisors with a grudge. All I had to do was show up and do my job. Bombs or not, I loved Afghanistan.

Free from the baggage I had back in the states, I quickly started talking to people and making new friends. *Is this what normal people do?* I wondered. *Cuz I really like this.*

Have you ever had a bad headache that wouldn't go away? That's how I felt in the States: constantly pent up, stressed out, kept down, and at war. But here, I flourished.

"So where are you from originally? Where are you stationed? How long have you got to go?" served as standard getting-to-know-you questions that we asked one another on deployment.

Some, on the other hand, cut right into the personal stuff, like, "Are you married? How's your relationship? Do you want to hook up?"

A new friend told me she planned to cut off all her hair.

"Your husband going to be okay with that?" I asked, not knowing I was about to get another lesson in the world of the normal.

"What's that?" she asked. "It's my hair. I'm the one that's got to be okay with it, not him."

"Are you Christians?" I asked.

"Yes," she answered. "So what? Jesus had hair. I'm sure he got haircuts," she responded, not tracking at all with why I asked.

The more I shared of my life and asked questions about hers, I realized that my marriage looked nothing like anyone's around me. When I told her about getting permission to cut my hair, receiving an allowance, and not having the right to say no when my husband demanded sex, her eyes grew wide and her eyebrows arched.

I tried to explain myself so she wouldn't think me a complete nut.

"See," I justified, "he doesn't think it's a good idea for me to have too much in savings, because that might tempt me to leave him." Even as I said the words, I knew the person saying them was, indeed, a complete nut.

The more I shared with other people and listened to their responses, one new friend summed things up in a brutally honest way.

"You know," she said, "federal prisoners have more rights and freedoms than you do."

"No," I said, defending SJ and myself. "I know it sounds bad, but…."

"It sounds bad, and I'm sure it's even worse. I'm not telling you what to do," she told me. "But girl, he's abusing you."

"No," I defended him. "He doesn't hit me." I decided it would be best not to bring up the time he tried to pull my teeth out.

"Domestic violence doesn't mean he gives you black eyes. It doesn't have to be physical," she shook her head. "It can be emotional and financial, too. You said he doesn't let you have friends. He manipulates you and your mom. He refused to let you get your own car, and you don't even know how to drive his car. He gives you an allowance, and if you piss him off, he deducts it from what he gives you of your own money. He spies on your phone calls and texts. He locks you in your house, and it sounds to me like he's raping you. So yeah, it is physical too. I'll say it again, you're in an abusive relationship."

After that, I didn't know what to do. If I kept talking, more people would think I was crazy. And the more they criticized and attacked me for staying with him, the more defensive I got. I could badmouth Ohio, my mom, and SJ, because they were linked to my heart, but no one else could rip on them. More than that, when it came to SJ, they weren't only judging him, they were judging me. Then a new worry popped into my mind.

What if it gets back to SJ that I've been talking about him? I gulped, adrenaline surging through me at the thought. *He would kill me for sure.*

Once that thought entered my head, I stopped talking with them about any part of my personal life. I spent the remainder of the deployment having superficial relationships and talking about safe topics, like bombs and ambushes.

While I was trying to pretend everything at home was fine, SJ grew increasingly angry with me for not immediately returning his calls. We spent most of our time on the phone with him asking me, "What are you doing? Why aren't you calling me back?"

"We've been in bunkers dodging bombs," I told him. "Every day, we come under fire!"

"Bullshit!" he answered.

"I'm telling you the truth," I protested. "After working twelve- to fourteen-hour shifts, I can't even go to sleep, because bombing keeps us in the bunker," I tried to explain.

He didn't believe me. Apparently, my deployment looked so different from his, that he demanded proof. When bombs dropped, we'd get incoming alerts telling us to seek cover. I had to take pictures of those alerts to send him as evidence.

Eventually, I stopped explaining or sending him proof. When he yelled at me, I would say the same thing each time:

"I'm sorry," I managed to say. "I'll try to do better."

Then I stopped communicating with him for a week. I felt like the walking dead in Afghanistan—exhausted from work, constant bombing, SJ woes, and lack of sleep. I wasn't trying to punish him, but I cut out doing the things that didn't keep me actively alive.

The first email he sent me during my communication sabbatical gave me some hope. "I feel like you're going to leave me," he wrote. "I feel like you're over there talking to someone else."

Then he said something I'd longed for but never saw a way it would come to pass: "If you want to get a divorce, get a divorce."

Yes, yes! I shouted in my head. *This is my time. Free at last, free at last...*

He initiated it this time, and all I did was concur.

My happy dance didn't last a minute before the next message arrived: "Good luck finding a place when you get back, especially with no money."

He had me, and he knew it. I couldn't set up finances from afar, nor could I switch funds or transfer direct deposit, because I had no individual account in my name to send it to. The only way I could leave SJ is if he agreed and granted me an equitable divorce, where I could walk away with what I'd contributed financially to the marriage.

I knew I was trapped for the time being, but my new friends had given me many things to think about. I knew my marriage was out of whack and that SJ held all the power—physically, emotionally, and financially. I began thinking about what I could do once I got home to give myself a touch of sanity. If I were trapped in this marriage, I needed some benefits to offset the abuse.

My deployment lasted seven months instead of the typical six. I loved my time away, but once I began the return to base, I knew I needed something to look forward to. That's when I developed a two-fold plan. First, I asked my mom if she would like to come live with us. She agreed. Second, I told SJ that I wanted to do something extra special for his birthday.

"I want to give you a nice surprise for your thirtieth birthday, so I'm going to need some money," I begged.

"How much do you need?" he asked.

"Well," I started, "I want to do something really big, so..."

While I'd been deployed, I earned around $30,000 or $40,000. Now, he gave me half of my earnings. I knew that he would never let me have money so I could buy my freedom or open an account of my own. So I figured I'd get something out of that money.

"Thank you, babe," I gushed. "It's all for your special birthday. It's all for you. I promise, it's not for me."

He took the other half and paid off his student loans. With my half, I planned a trip for us to Tahiti. What did I care how he spent my money, as long as I could spend at least some of it on something I really wanted to do.

Tahiti turned into the best time we had together in our entire relationship. First, we were away from my job, so I had zero stress from my supervisor. Second, SJ relaxed and didn't feel the need to control everything—because I controlled everything for him. For example, this was the only time he allowed me to wear a two-piece bathing suit on the beach. Why? I got us a private villa with a private beach and pool. And I didn't get just *a* bathing suit. I got ten! I rotated them out every two hours. Every time I went into the pool, I put on a new one. I felt so normal, wearing clothes I liked and not being slut-shamed in the process. This was the life!

We returned to Virginia in December, and my mom moved in with us the next month. I'd stacked the good things in my life to happen in the longest, most unbroken row I'd ever had. SJ didn't argue about my mom moving in. He liked her at the time, especially when she sided with him in any arguments the two of us had. Also, he took compassion on her. Her health wasn't great, and the free clinic where she worked had closed its doors.

Then came another bonus: SJ wouldn't fight with me in front of my mom. We experienced our longest period of peace since we'd been together.

It wouldn't last. Once SJ saw my mom and me getting close again, he started resenting her. Eventually, he talked down to her much the same way he mistreated me.

I became close with a dear woman, a retired E9 who worked as the coordinator of Sexual Assault Prevention and Response (SAPR). When I felt overwhelmed by the abuse or harassment at work, I used her as a sounding board. I met her as a result of the Morin ordeal, and we stayed close after that.

When I returned from deployment, she excitedly told me of a regulation change that took place while I was gone. Any victim of sexual harassment whose perpetrator was found guilty of an Article 120 (rape and sexual assault) qualified for expedited transfer to a new base. Prior to that, only rape victims could put in for expedited transfer.

"Say the word, Kia," she said with a twinkle in her eyes, "and I can help you move out of here. You'd have a fresh start!"

Her words and kindness gave me something to hang onto. She also gave me some advice that I practice even today.

"Document everything," she warned me. "Every time someone says something or does something to you, write it down. That can serve as evidence about the ongoing chain of events you've endured. We can take everything you document to your lawyer."

With her help and the guidance of my newly appointed Special Victims' Counsel lawyer, I completed the paperwork and selected three bases that SJ was willing to transfer to as the best options for the Air Force and our family.

It's rare that plans work out as intended. But this one worked as designed. Boom! We got nearly instant approval to relocate, and we planned our transfer out to a sunny, beautiful base in Florida. Before leaving for Florida, we found a house online that looked ideal, one large enough for me, our dog Bonsai, SJ, my mom, my two sisters, and my nephew. Even though SJ didn't like the idea of having so much of my family under one roof, he loved the idea of charging them rent. While Mom and I packed up our stuff from Virginia, my sisters and nephew packed up their things in Ohio, and we all met in Florida to move in.

Unfortunately, we had to wait before we could get into our house, so we checked into a hotel for a short stay with my mom, SJ, Bonsai, and me. When my sisters and nephew arrived, we had six humans staying in one hotel room with two beds. We knew it would be for just a short time, but things got heated fast when we were cramped together. Within days, SJ started yelling at everyone, calling my mom and sister "a bunch of moochers" and "gold-diggers." Never mind the fact that half of that "gold" came from my earnings!

Finally, we got the okay to move into our house. Having extra space gave us a little more distance from each other, which helped some. SJ seemed happy about the benefits of my mom and sisters helping him pay the mortgage, and he relished using them for free labor, charging them with cooking and cleaning. But this kept him happy for only so long before his core nature returned.

SJ instituted new rules where he played the dad of the household. Rule one, take off your shoes once you enter the home. Rule two, don't touch the walls. Rule three, if you touch the walls, you must scrub the walls. Soon, he instituted curfew and even a prohibition on the hours anyone could open the refrigerator door. Finally, any guest we wished to enter the home had to be cleared by him in advance. Before he'd grant permission for guests to enter, he needed to know their entire backstory, how long they would be staying, and what areas of the house they planned to enter.

It wasn't long before he started conducting weekly room inspections. Mind you, not just for the child (my nephew, who was ten), but the grown-ass adults, too. If a bedroom "failed" his inspection, he'd give the occupant a certain amount of time to clean it up. But the clock was ticking. He'd warn them that if the room weren't spotless when he returned, he'd throw them out.

So, I guess you could say, my mom and sister got a little firsthand experience of what my life had become. My nephew left a few marks on the wall as boys do, so SJ made him scrub the wall and repaint the area. It didn't match perfectly, but the punishment wasn't about how the wall looked when repainted. It was SJ's way of saying that he alone ruled this castle.

Once we got settled, I determined to make good at my new base. Free from the reputation my old base had branded me with, I threw myself into work to make a great first impression.

I soon learned that I could change supervisors, bases, and states, but some things would never change.

Chapter 20

New Base, Same Old Shit

"Things are more like they are now than they have ever been."

~ 1948 classified ad

I FOUND LOTS to love about Florida. Here I had family, sunshine, beaches, and even a friend from Basic. My only complaint is that I didn't get here sooner!

It's customary when you transfer to a new base to be appointed a sponsor. My sponsor was a staff sergeant named Kramer. His job was to take me on a tour of the base, show me my work area, and help me acclimate to the new digs. He did a great job on the warm fuzzies and welcoming me to the base.

On my first actual day of work, he showed me the armory.

"This here is my baby," S. Sgt. Kramer said, spreading his arms to show off the large building. Once we got inside, he locked the door. "To me, this place is more than where I work. I come here for me-time. No one can get in here without me letting them in, and no one can get out of here unless I let them out."

I totally understood about having a place for me-time. But that whole thing about no one getting in or out sounded a little…off. *But what do I know? I'm new here,* I thought as I tried to let go of those comments.

"Yeah," he continued. "This place is 100 percent secure. Like I can play music as loud as I want. Nobody can hear it. I can scream in here, but nobody outside can hear a thing."

Yeah, the hairs on my neck started to spike, *that's not a creepy thing to say at all. No, nothing creepy about that.*

Alarms rang in my ears telling me to get the hell out of the armory and away from this guy.

"You know," he said, "you look real good." He started licking his lips as he eyed me up and down.

"Stop," I said. "I feel uncomfortable. I want to leave now."

"No, no, no, no," he said, giving me the settle down hand gesture. "Don't be like that. You got a cute frame on you."

"Thank you, but no thanks," I said flatly. "I want to leave now."

I tried pushing past him to the door.

"No, no, no, no, no," Kramer said with his hands up. "Don't be like that, don't be like that. I'm just playing with you. Come on. Let's count these weapons."

"No," I cut him off. "I don't want to be in here. I'm extremely uncomfortable."

"Okay then," he said. "That's fine. We'll come back later."

He headed to the door and then turned around to stare at me. I backed up so I was a few feet away. He continued to look me up and down.

We were in a room full of weapons, all locked, with no ammunition. He had an M9 on his hip, and it was loaded.

After a long time of him looking me over, he finally unlocked the door. I bolted out immediately.

In all armories, protocol states that you have two parties check out. I didn't care. I didn't bother. I ran out and didn't look back.

My God, I thought as I hustled away. *Was this dude going to try to rape me inside the fucking armory? Why else would he let me know he controls the entrance and exit, and why would he tell me the building was scream-proof?*

I rushed to my car and called SJ. I told him what had just happened.

"Did he touch you?" he asked.

"No," I said. "Well, yeah, he touched my hand, but no, he didn't grope me or anything. But I'm still shaking."

"Look," he said calmly. "We just got here. Maybe he didn't mean anything by it. Just put him on your radar, okay? Don't say anything. We just left a place where you were assaulted. You're still raw right now. Just ignore it," he suggested.

After thinking about it, I decided that SJ might be right. Maybe I was a little raw and jaded from past experiences. I chose to let it go. But I kept in the back of my mind never to go to the armory with him again.

So that's what I did. The next day when Kramer approached me, I normalized things. After all, he had been assigned to in-process me to the base, and part of that meant he drove me wherever we needed to go at work. I thought nothing about it when in the middle of out-processing me, he drove off the base.

"Where are we going?" I asked.

"I'm taking you to lunch," he announced.

I reminded myself: *It's time to make a fresh start and put those nightmares behind you.* But that didn't mean I was going to be anywhere alone with him, either.

"Okay, sure," I answered. "Where are we going to eat? Because I'd like to invite my husband."

"I don't know the name of it," he shrugged. "It's just a burger joint."

I texted SJ, suggesting he look up nearby burger places because Kramer told me he didn't know the name of the place.

As we pulled up to the restaurant, I texted SJ the name of the place.

"My husband is on the way," I told Kramer as we walked inside.

"Why are you telling your husband where we are going?" he asked as if I'd offended him.

"I don't know," I responded. "Maybe my husband is weird, but he likes to eat, too."

"Well, that's fine," he said in a huff, "but I ain't paying for him."

"Nobody asked you to pay," I snapped back. "I can pay for myself, thank you."

"No, no," he shook his head. "When you're with me, I take care of everything. I got it. I took you out; I got this."

Figuring SJ was en route, I just shrugged and changed the subject. But SJ never showed up. He told me later that he couldn't find it.

At the end of the day when SJ picked me up, we talked over what had happened at lunch.

"There's something not right about Kramer," I told him. "He gives me the creeps. I don't want to be alone with this guy anymore."

A month later, I worked in the office with Kramer while everyone else took lunch. An airman sat at the front of the office taking care of customers, and Kramer and I sat at the desks we'd been assigned in the back of the office outside the office of the NCO, who also had left for lunch.

Kramer kept making strange grunting and moaning noises. Annoyed, I found something to do at the front of the office just to get away from his noises. When I returned to my desk, Kramer was gone.

At the back of the warehouse, some airmen used a weight bench and free weights to work out on during breaks. Kramer walked into the office from the direction of the weight bench, and he was dripping with sweat.

"Hey, Chitty-Chitty Bang-Bang" he said to me, a nickname he assigned to me that I loathed. "Can you spot me?"

"You want me to spot you?" I asked, shaking my head. "I've seen how much weight is on that bar. So, um, no!"

"Na, I'm just kidding," he said, lifting his nasty, sweat-soaked shirt to wipe his face. "I just put in a good workout."

"Man," he continued. "I'm really sweaty and hot. Do you want to feel me?"

"No, not at all," I answered, looking away.

"No, it's alright," he insisted, "Come here, come here. Feel me."

Kramer creeped me out so badly that I kept pepper spray on my keychain. I subtly reached into my pocket to make sure it was there.

"Come on, now," he persisted. "You're so little I bet I can bench press you."

"Stop, please," I said nicely but firmly.

He walked towards me, and I stood up. He kept coming at me, and he literally backed me into the corner. The desk, water cooler, and wall were on three sides of me, and his sweaty body touched mine. He groped me as I screamed.

I stuck my hand in my pocket, but the deep pockets of the cargo pants made reaching the pepper spray impossible. So instead of continuing to fumble for it, I continued to scream while pushing him away from me.

"Get the fuck off me!" I repeatedly cried out as I broke away from him and ran out of the building.

In tears, I raced to SJ's job, sobbing so hard I struggled to keep the car on the road. I was grateful that he let me drive his new car, since I had a day of appointments and didn't want to be stuck in a vehicle with Kramer driving me anywhere. After telling SJ what happened, he got in the car, drove us back to my job, and confronted Kramer.

"Who the fuck do you think you are putting your hands on my wife?" SJ shouted.

"Whoa, brother," Kramer said defensively. "Calm down here. I don't know what she told you, but you got to calm down."

The airman that had been working the customer desk heard the shouting and came into the office where the three of us were yelling. Then he just stood there.

Where the hell was this guy when I screamed? I wondered.

I'm crying uncontrollably while the other three men look at me. Then Kramer jumps in.

"Like I said, I don't know what she told you, but I'm so sorry whatever happened," he said.

"Whatever happened?" SJ repeated sarcastically. "My wife said you threw yourself on her, and you felt her up."

"Oh, that?" Kramer responded. "I didn't mean nothing by it," he continued. "See, I just got done working out, and I wanted her to feel how sweaty I was."

As if somehow that made the situation okay, like he was expecting SJ to say, "Oh, I'm sorry I overreacted. I get it. Your hands were wet with sweat, and you just wanted to dry them off on my wife's breasts. No harm, no foul. I dry my hands on the closest set of knockers I see when I'm sweaty, too. We're good."

But SJ didn't say that. He said nothing. SJ acted like your prototypical alpha male; but in reality, he only bared his teeth to females, meaning me.

Kramer kept going. "Listen, I'm a man of God," he said, repeating lines I'd heard too many times by other abusive men, including SJ. "I promise you I didn't mean nothing by it. I put my right hand on the Bible. I promise you I meant nothing by it. I mean, if someone did that to my wife, I would be upset, too; so I understand where you're coming from, brother, I understand. But I promise you, it was..."

He stopped in mid-sentence and turned to me.

"Kia, I'm sorry," he said.

"Don't address her. You've lost the right to talk to her," SJ told Kramer.

"Well, I am sorry," Kramer continued. "If I hurt you, or if you felt violated, I apologize. I did not mean it at all. Right here, right now, under God's word, I meant no ill intent."

Then he appealed to SJ.

"You're a man of faith, aren't you?"

"I am," SJ said, his face showing no emotion.

"I could tell you were a man of faith," Kramer responded. "Right here, you're practicing Matthew 18, where Jesus says, 'If your brother commits an offense, go to that brother, and point out his fault, just between the two of you. If they listen to you, you have won them over.' That's what you did right here. And I thank you for it. You don't need to do anything else, I promise. This will never happen again, my brother."

SJ kept his hands on his hips, and he looked over at me.

"What do you want to do?" he asked me.

Part of me had an ounce of gratitude that SJ took the time to listen to me, come down there, and try to resolve the problem. But a much bigger part of me thought:

What in the hell? I didn't ask you to come down here, wave your hands, and then turn it back to me to decide what to do.

"Let's just go," I said, shaking my head. Then I headed to the car without answering SJ's question.

As we were walking out, Kramer ran after us.

"Hold up, hold up," he said. "Before you guys go, I need to let you know I'm just going to have to call my boss to let him know you came down here—just in case, you know, you escalate this little issue. I mean, I can't let my people be blindsided that you came down here and started going off on me. So I have to let them know."

"Fine," SJ answered. "And I'll let my people know, too."

Instead of coming down to my work and defending me—and I'm not talking about a fight to the death—SJ basically did nothing. As a result of this episode, Kramer reported the incident to his boss, and SJ ended up getting a no contact order, meaning he was barred from visiting my workplace again.

I took matters into my own hands and reached out to the Special Victims Counsel I had at Langley AFB. Since she had not yet closed my case since my expedited transfer to Florida, she extended her services to me and notified my chain of command. That initiated a very quick investigation.

In the end, Kramer got a letter of reprimand (LOR), which given the facts of his actions, served as a slap on the wrist. I had gotten a letter of counseling for responding to a text from an NCO at work who then exposed himself to me.

After this wrapped up, I told SJ, "The next time someone invites me to Bible study, I'm just going to tell him to drop dead."

All of this happened within three months of my arrival at the base, which reminded me too much of what had happened to me at Langley. Another thing Kramer had in common with S.Sgt. Morin was that both men were loved in the squadron. And again, like S.Sgt. Morin, Kramer was tight with his master sergeant, who immediately spread gossip about me, saying, "She's been here three months, and this shit already? She's going to be problems, coming down here thinking she's too damn cute. No, she's going to be problems." Then the master sergeant reached out to a colleague at Langley who happily shared all the details of my time there.

S.Sgt. Morin and Kramer had one more thing in common: I was not their first victim. When I told a coworker about how much Kramer scared me, she confided something in me.

"I'm gonna tell you something," she said. "Before you got here, he used to treat me that way. He used to be all over me. He used to make me feel so uncomfortable, always making comments about my butt and my lips and everything."

"So I'm just his latest toy to mess with?" I asked incredulously.

"All I'm saying is that once you got here, he left me alone," she concluded.

Most airmen in the Air Force are good, decent human beings. But the bad ones? People are afraid to say anything about them. If you talk about what happened to you, you can end up like me: vilified, smeared, harassed, and hazed. But when no one says anything, more people fall victim.

Yep, I learned that some things never change.

But some things do. I learned I was pregnant.

Chapter 21

Demon Dolls and Baby Blue

"You are proof that love before first sight does exist."

~ Araceli M. Ream

WHEN I LEARNED I was pregnant, I experienced two opposite emotions.

First, I found myself overjoyed about being a mom. I had an amazing mom, and I couldn't wait to have a child of my own to mold, shape, and teach as my mom had done for me.

Second, I lived in dread—because of SJ. He didn't want to have children, saying he didn't feel ready to be a father. His daily emotional ups and downs scared me to death.

Interestingly though, even though he eventually got me pregnant, he had started to find me "physically repulsive." He had so much jealousy that he threatened to kill any male that talked to me, and he repeatedly took advantage of me sexually a year earlier. But once my weight "spiraled out of control," we'd had sex twice, and one of those conjugal visits resulted in my current state.

The reason SJ did not find me attractive is that I had "ballooned" to 116 pounds. To give you an idea of how *obese* I was, the ideal weight for a 5'5" female is approximately 136 pounds. SJ had such an obsession with his idealized, perfect woman frame that being twenty pounds *underweight* wasn't skin and bones enough for him.

Once I got pregnant, I became even more repellant to him. I didn't complain. I liked not having to be on-demand for his sexual desires and abuses.

SJ displayed bipolar behaviors throughout our relationship, but once he learned a baby was on the way, his mood swings amplified. Some moments he would love me and say, "You're carrying inside of you this beautiful creature." Other times, he'd say things like, "You're the one who wants this baby, so don't expect anything from me."

I quickly learned that when SJ came with me to my ultrasound appointments, things became real and wonderful for him. But as soon as the image of that small child inside me no longer remained fresh in his vision, he would return to being an ass. Due to the frequency of those ultrasound appointments, we fell into a two-week cycle. One week, he would show excitement, and the next week, he'd forget all about the baby. We continued on this track throughout my pregnancy.

I had a hard pregnancy. I was diagnosed with hyperemesis gravidarum, which is like an extreme version of morning sickness. To give you an idea of the severity, I would puke about ten times each hour. Sip of water: puke. Smell of bacon: puke. Nothing in my stomach: puke. I threw up so constantly that I had a standing appointment at the ER to receive intravenous fluids to prevent dehydration and malnourishment.

I ended up on convalescent leave, extending from January 31 until I gave birth to my daughter, because of those complications. While I was carrying her, she kicked one of my disks out of my back that affected my legs. Between puking, delivery, and limping, I was out of work for nearly ten months.

A couple of months before my daughter was born, my sister offered to do a maternity photoshoot of me to capture pregnancy in its glory. She took multiple photos of me wearing high heels and leaning against a large tree. Since I couldn't walk in those shoes, I took them off and carried them when we moved from one location to the next. In the process, a strap fell off my shoe and I couldn't find it.

"SJ, the strap fell off my shoe," I said after looking around the ground. "Do you mind going to get it?"

"Yes, I mind," he answered as if I'd just asked him to swim through a sewer. "If you want it that bad, go get it yourself."

"I'm tired, and my feet really hurt," I pleaded.

"Oh, well," he answered.

He wouldn't go get it for me. That was the SJ I knew, starting from three weeks after we started dating. He lived for himself, and he could be ornery and mean for no reason.

Closer to my due date, I started having contractions. We rushed to the hospital, and I learned that my dilations had stopped. Instead of admitting me, they sent me home with Ambien to help me sleep. By the time I got home, I felt like I'd taken acid. The Ambien messed up my mind and created vivid, disturbing visions and dreams.

Looking over at the crib in our bedroom where we had a baby doll waiting for her, I watched as the doll came to life and started screaming in a demonic voice. My room turned into a horror movie. The doll ran across the room, shrieking in a dreadful tone. In my mind, that baby doll was Annabelle or the Bride of Chuckie! I freaked out.

"Be quiet," SJ offered instead of comfort or compassion. "Shut up and go to sleep!"

"Help me, please!" I pleaded. "She's staring at me! She wants my soul, and she's going to hurt the baby!"

I've never had night terrors, and this drug-induced hallucination is as close as I want to get to them.

Finally, I fell into a deep sleep for a couple of hours. When I got up to use the restroom, I noticed that I was soaked.

"I'm so sorry," I said to SJ. "My water broke in the bed."

SJ turned furious.

"In bed?" he shouted. "Why are you just fucking standing there. Help me strip the bed and go get new sheets. Then clean up the floor."

Going into labor, I didn't have time to worry about the baby or the delivery. Instead, I'm putting clean sheets on the bed and cleaning up the floor.

Once I finished, I called out to my mom upstairs to tell her the baby was on its way. Mom got so excited she ran down with her camera to take a final *before* picture. My back hurt like I had a knife sticking out of it, and I was doubled over in pain. But when I look at that photo, what stands out the most to me is the disgusted look on SJ's face. His face said, "For God's sake, will you stop bitching and moaning about every little thing! Stop making this all about you!"

If I didn't have that photo, I'd have little recollection of what happened that night. The Ambien kept me foggy and out of it. I don't remember getting in the car, checking in at the hospital, or getting into the hospital bed. But I do remember lying in the hospital bed with an IV ready for an epidural.

As a side note, I didn't want an epidural. I had been quite clear about this to my doctor and to SJ. However, once I started screaming in pain, SJ overruled my decision. He had me sign off on it while I was semi-conscious. The next time I became coherent, they were prepping me for the epidural. When I asked him later why he did that to me, he said, "You were too loud. I needed you to be quiet. And you were in pain. I thought it would help you settle down."

I was still coming down from Ambien, so I had no idea what these nurses were telling me. Even though they'd tried to explain it to me, I wasn't fully aware of what was going on around me. SJ told me I was signing a consent form to get admitted into the hospital. Moments later I was being prepped for a bedside procedure for the epidural while I was still out of it. I don't even remember the pain.

While still in active labor, I wanted to text my family to see if they would like to come visit me. SJ refused to let them come, going so far as to put my phone on a counter across the room so I couldn't reach it. After I had pushed for a few hours, they took me in for an emergency C-section after the baby swallowed meconium. Her condition, coupled with other complications, lowered both of our heart rates to a dangerous level, and the doctors worried that both of us might die on the table.

While I already embraced motherhood, this moment cemented it for me.

"Save the baby," I told the doctors. "Don't worry about me. You can let me die, but take care of my baby."

In no time, they strapped me to the operating table. A combination of Ambien and whatever else they put in my body made me twitch and convulse, and they didn't want me moving as they cut me open. Feeling nothing except fear, they swiftly cut me open and took my baby out.

Something was wrong. The doctors were speaking all at the same time. But I didn't hear any crying. I strained my neck to see her. The tiny body the doctors huddled around looked lifeless and blue.

"What's going on?" I shouted. "She's not moving!" I cried as my mental state swung into full panic.

I watched as they performed CPR on her while suctioning her nose and throat.

I heard the doctors ask SJ if he wanted to see his baby.

Oh my God, my baby is dead, I thought as I started to have trouble breathing and my vision turned dark.

"What's going on?" I repeated. "My baby!"

SJ stood somberly while he looked down at our baby for the first time.

The next thing I heard might as well have been the "Hallelujah Chorus" sung by a choir of angels. I heard my baby cry!

"Let me see my baby!" I screamed. In response, my baby cried louder.

"Please, bring her to me. Let me see my little girl," I begged.

Eventually, the doctor brought me my sweet little girl and placed her against my face.

"Here's your baby girl," he smiled. "She's beautiful. Congratulations."

"Reign," I said. "Her name is Reign," I repeated as I looked at her perfect little face.

Instantly, Reign stopped crying and tried to suck on my cheek.

Within seconds, the doctors took her from me and rushed out of the room.

"Reign!" I screamed. "Where are you taking my baby?"

I tried to get up from the table, but they had me strapped down like something from *The Exorcist*, and I couldn't move.

"She's being transferred to NICU," a nurse told me. "She's going to be okay, honey. We're going to take good care of her. Now we need to take good care of you."

SJ left without saying a word. They wheeled me into recovery for an hour that felt more like a lifetime. Finally, they wheeled me into NICU, where I could see my precious Reign with cords and wires hooked up all over her tiny body. Then I saw SJ standing next to her. I learned that he insisted to be with his baby girl and didn't want to leave her side. What a pleasant surprise.

Once the drugs left my system, I took a nap and then texted my family.

Mom responded right away.

"You had the baby?" she wrote.

"Yes!" I replied. "Didn't SJ tell you?"

I looked over at my husband, who stood there with a blank look.

"Oh, yeah," he said. "I forgot to tell them. But I told my mom."

He texted his mom in Korea about the baby, but he didn't text my own mom who I wanted to be there with (and for) me throughout the delivery! Reign had been born for seven hours before they even knew she arrived, and they knew nothing about how we both nearly died on the table.

All throughout my pregnancy, SJ would not allow members of my family to come to doctors' visits. He said to me, "Did they create this baby? No. So why do they need to be here? Why do they need to be a part of it? This has nothing to do with them."

Reign was in the NICU for almost four days. The first time I got to hold her for more than a few seconds was on the third day. That's when SJ allowed my family to come up to the hospital. But he had some stipulations.

"No pictures," he told my family as he held up one finger to indicate more rules were coming. "No asking questions," he said as another finger appeared. "You may hold the baby but for a maximum of five minutes before you give it to the next person. Then she comes back to me or her mother," he finished as a third finger popped up.

My sister Brandy wasn't having any of that.

"I got a finger of my own I want to show you," she said. "Just give me that baby."

From the first day in the hospital, I felt completely drained of strength, and fatigue hit me hard.

"Would you like an Ambien to help you sleep?" a nurse offered me one night.

"Sure," I answered, completely blacking out the experience I'd had just a day earlier.

Once again, those chemicals scrambled my mind. As the medicine entered my brain, I looked up and saw an intruder sitting in my hospital room. Fear gripped me as the intruder took on the appearance of the grim reaper.

"Get out of here!" I shouted while trying to push myself to the farthest side of the bed. "You can't take me or Reign! Get out!"

Looking around the room, I saw my phone. I picked it up and called my mom.

"Don't you see him?" I shouted to her. "He's an invader. I don't know this man. I want him out! Please don't let him hurt me!"

Mom's heart broke as she heard the horror in my voice. She took on my fear as her own, and she called SJ to plead with him to help me.

"SJ, will you please try to calm her down?" my mom asked him. "She's just had a baby. She's under so much stress, and whatever medicine they gave her is messing with her mind."

"No," he told her. "I'm going to get the doctor. She'll be fine, but I need to get some sleep."

SJ rang for a nurse.

"Can you give her IV fluids to flush that medicine out of her system?" he asked reasonably. "She's having a bad reaction to Ambien, and I can't sleep."

After that night, he told the nurses not to give me anything to help me sleep.

Coming home from the hospital, SJ took Reign from the car and went into the house, leaving me sitting in the back seat. The C-section had made it difficult for me to swing my legs out of the car, but he didn't care. SJ hadn't wanted to be a father. Now he did. His feelings vacillated throughout my entire pregnancy, but once Reign arrived, it was all about her. To him, I had been useful for carrying his child, and now I became a wet nurse and diaper changer.

I hoped maternity leave would allow me to recharge physically and emotionally as well as give me time to bond with Reign.

I got a break, but it wasn't the one I expected.

Chapter 22

The End of the Beginning

"There are wounds that never show on the body that are deeper and more hurtful than anything that bleeds."

~ Laurell K. Hamilton

SHORTLY AFTER GIVING birth, I suffered from the baby blues. I believe I had postpartum depression, but I left out a lot of information when I met with the doctor because I feared receiving that diagnosis. Since I remained in active duty, a postpartum depression diagnosis would have sidelined me and brought too much unwanted attention my way. Sheesh! I had enough of that already.

We had water overflow from a river behind our house that fed into a bay. Wherever you have fresh water in Florida, you're going to have alligators, snakes, and other creatures of the *oh, hell no* variety. Whatever Florida has, it was running up and down my back yard, and you could see it all the time.

One day when my depression peaked to meet with SJ's wrath, I snapped. While I was accustomed to his verbal abuse, it came on top of exhaustion from not being able to sleep and constant commands to "pump more," "feed the baby," "give her a bath," and "change her diaper." I had become his full-time maid, expected not only to care for his every need, but also to take his direction on everything Reign needed.

After putting Reign down for a nap, my body and mind felt numb. I walked out to the water behind our house, crying without realizing it. I knew the water was full of alligators. I'd seen them swimming back and forth all the time. While I didn't see any at the moment, I wasn't looking either. I knew that as soon as I hit the water, they would appear. I kept inching my way to the edge of the overflowed river.

"I'd like to swim with them," I thought aloud. "I can join them for a little water time."

I stepped closer and closer to the water's edge. As my foot was about to enter, screaming broke my trance. My face still dripping with tears, an alligator had gotten within a couple of feet from where I stood without me noticing it.

"Kia, get back in here!" SJ yelled from the back door. "Reign's awake and crying. Come take care of her. She filled her diaper."

Mindlessly, I returned to the house, changed her, and put her back down to nap.

Meanwhile, SJ didn't ask what I was doing or if I was okay. He didn't question why my face was swollen and tear-stained.

When Reign turned two months old, my mom broke one of SJ's cardinal rules: she held Reign for too long. This infraction escalated into a huge argument where SJ lit into my mom.

"You don't obey my rules in my own damn house," he shouted at her. "I don't know why you're even living here. You don't work. You don't contribute financially. You're always saying you're tired, but I don't know what you could possibly be tired from, because you never do anything!" He encapsulated everything he could hurl at her in one succinct phrase.

During his litany of insults, I stood in the middle of the two them, holding Reign.

"You were wrong," I finally said to SJ. "Don't talk to my mom like that. You're just being cruel. What the hell is wrong with you?"

"Shut up, and stay out of it," he snarled back.

I looked at my precious baby, then I glanced back upstairs. I couldn't win. I ended up taking Reign into my bedroom so the two of us could get some peace. But that peace didn't last long. SJ came in, red-faced, and began to punch the wall. I wrapped my hands around Reign's ears and rocked her.

Eventually, SJ calmed down, and when he did, I decided to do my best to smooth things over between Mom and SJ.

"Hey, do you mind if I bring Reign up to see Mom?" I asked. "Today's been hard on all of us," I said, not really sure how holding a baby for "too long" created all of this drama, "and Reign always cheers Mom up. I just want to make sure she's okay."

"Fine," SJ answered without looking up. I'd long known that in lieu of an apology, SJ would "make up" by not screaming at me or arguing with me.

"Five minutes," he said, as I headed to see my mom. "I'm setting the timer."

"Knock, knock," I said softly outside my mom's room.

She didn't reply.

I opened the door a crack, and I could see her on her bed with her eyes closed.

"Are you taking a nap, Mom?" I asked. "Because I brought you your favorite person. Reign's here to see you."

Mom didn't respond or stir.

"It's okay," I told her as I crossed the room and sat on her bed. SJ said you could hold her now."

"No, I don't want to hold her right now," she slurred.

"No, it's okay," I said, placing Reign in her arms.

Instead of Mom holding the baby, Reign rolled out of her arms. When I touched Mom's arm, she felt cold and limp.

"Mom, wake up," I said, placing Reign in the middle of the bed. "Wake up, Mom!"

I pushed her to get her to move. Her limp body rolled back to its original position after each push I gave her, like she was deadweight.

I freaked.

"Alexis!" I screamed for my sister. "Alexis, get in here! It's Mom!"

"What's going on?" Alexis rushed from across the hall to Mom's room.

"I don't know!" I wailed in a high-pitched voice. "She won't wake up!"

"SJ!" I screamed as loudly as I could. "Please come help us!"

At this point, Alexis and I were trying to get Mom to wake up, and Reign was screaming from all the sudden noise.

"What?" SJ said coldly and flatly from the doorway.

"Call 911!" I shouted. "Something's happened to Mom. She's not waking up!"

Next to me, Alexis kept gently slapping Mom's face to try to keep her eyes from closing.

"I'm not calling anybody," SJ said with disgust. "She's just doing this shit for attention. There's nothing wrong with her."

With that, SJ leaned up against the doorway and just watched.

"What the hell?" Alexis shouted. "Fuck it! I'll call."

"Hey!" SJ shouted across the hall. "I don't want all this drama in my house!"

Alexis ignored him and placed the call. Within minutes, an ambulance pulled in front of the house. They put Mom on the floor, and then they started doing a sternum rub on her. Finally, she woke up a bit.

"Any idea what happened here?" a paramedic asked.

We had no idea, but my mom's eyes went to the bathroom. The paramedic saw her glance, and he went inside the bathroom, turning on the light. A second later, he came out holding a bottle in his hand.

"Did you take these?" he asked, holding up the bottle in front of Mom's face. In response, Mom looked away.

"Let's get her on the stretcher," one medic directed another. "We need to get her to the hospital."

Alexis drove her own car to the hospital, and SJ drove us in his car. We sat in the waiting room for hours before they let us see her. By the time we got to her room, Mom was drinking activated charcoal to keep the medicine she took from being absorbed in her gut. Alexis and I were weeping, scared for our mom. But I cried, too, knowing SJ had pushed her to this dangerous brink.

Meanwhile, SJ stood at the furthest point of the room, glaring at my mom.

"She does this shit on purpose," he muttered.

"Not the time or place," I said, trying to make sure things didn't explode again.

Once Mom was out of danger, we all went home for the night.

The next morning, SJ stood waiting in the kitchen for me.

"She's got to go," he announced. "She's not welcome here any longer."

"Are you serious right now?"

"I can't be around all this negative energy," he said sternly. "I can't have this shit in my house."

He can't be around this negative energy? Who does he think is, creating all this negative energy?

"My mom literally just tried to kill herself because of you," I yelled at him. "And you want to kick her out?"

"Yup," he answered.

"Do you know how screwed up that is?" I challenged him. "You pushed my mom to the point of wanting to be dead rather than to live in this chamber of horrors you've created, but when she tries to kill herself, that's where you draw the line?"

"Whatever," he said without emotion. "Either you tell Alexis and your mom that they're moving out, or you can move out with them. I don't care. But Reign and I are staying right here."

Brandy never moved in with us at the house. She had enough of batshit-crazy SJ at the hotel when he screamed at everyone. She found a job and got a place of her own. Alexis had been saving her money and looking, too. And with SJ's ultimatum, I needed both Alexis and Mom to leave.

Brandy lit into me over this decision.

"Wait a minute!" she snapped at me. "Mom tries to kill herself because of your crazy-ass husband, and now she needs a place to recover from a situation that you were part of? And you're throwing her out? You're seriously choosing your husband over your mom? Look what just happened. Are you going to stick with him after he kills us all? Are you freaking kidding me right now?"

I couldn't keep Mom, Alexis, and Brandy from seeing SJ screaming and punching holes in the wall. But I never let them know about all the other things that went on out of sight. They had no idea that he sexually assaulted me, locked me in my own house, put me on an allowance, threatened to kill my friends, or a litany of other things.

They were pissed off at me, and they cut me off. I don't blame them. They didn't know what I didn't tell them, so they drew the only conclusion left to them: I chose SJ over them.

By this time, Alexis had found a place of her own, and she was just waiting for the move-in date. She had been planning on moving out anyway, but she didn't want to say anything beforehand, fearing it would escalate tensions with SJ and the rest of us. But this cemented her plan of action. On her move-in date, she and Mom left.

I had been lonely and isolated before, but this was a new low. I lived in a five-bedroom house that felt cold, sterile, and empty. After they moved out, Mom recovered, even getting a job and being happier than I'd seen her in a long time. They all did better emotionally once they got out from under the toxic cloud of my home.

I never knew misery like what I experienced at that point in my life. And I knew that I couldn't go on much longer.

My friend Vickie threw me a baby shower, and she flew down from Virginia to make it happen. When we were alone together, I poured out my heart to her.

"I can't do this anymore," I told her. "I have to divorce him."

"Yeah," she agreed. "You're in way too deep. You've got a house and baby now, too. I don't know how you're going to get out of this without a fight. You'd have to sell your house. And if you want custody, you need your own money. You can't even pay for a lawyer right now, the way he has your money locked down."

Her words pushed me in the right direction. I needed to sell the house. But how?

"Hey, SJ," I said during a calm part of our cycle, "since we don't have my mom or sisters here, this is way too much house for us, and it's getting harder to pay the bills without them chipping in. Why don't we try to sell the house and turn a profit?"

Surprisingly, SJ didn't disagree.

"Okay," he nodded. "I'll see what this place is worth." Then he posted on a realtor site, and an interested party reached out to him immediately.

"Let's ditch the realtor," Derek, a potential buyer, suggested, "and then we can both save money?"

SJ agreed. It was a disaster. The buyer was a real-life alpha male, and he controlled SJ the same way SJ controlled me. All the terms favored the buyer. Not only did we enter a rent-to-own deal with the buyer to take our home, but we ended up renting a place from the buyer. Essentially, we swapped houses.

Long story short, Derek screwed us over. We ended up staying in our home for seven more months than we wanted to, wasting seven months of money I could have tried to save to put fuel in my escape pod.

One good thing came of it, though. SJ became so increasingly cruel and ugly to me during those months, I knew that he would either kill me, or I would get out. At that point, I didn't care which. But I was sure of one thing: this madness needed to come to an end.

Chapter 23

Making Rank and Stirring Pots

"You pay for good days by then having bad days. You pay for joy with pain."

~ Taylor Swift

I RETURNED TO work, still suffering from what I called the "baby blues." But my return coincided with a major accomplishment: I made rank as an E5, a non-commission officer. After my promotions, I attended Airman Leadership School, where I ranked in the top 10 percent of my class. From the entire class, I was nominated for the Commandant Award in addition to bringing home the Distinguished Graduate Award to my unit. These were top honors. My squadron hadn't had a member win either of those honors or any other awards for a long time.

As my dubious reward, they gave me a troubled airman, Karen, as a direct report. This airman had already been talked to by an E8 regarding perception: she kept leaving our building to talk with a married NCO. When my supervisor assigned this airman to me, he let me know that a conversation had already taken place. Thank goodness, I didn't have to deal with that. Instead, I wanted to get to know her and see if there was anything I could do to help.

"Hi," I started. "How are you doing?"

"Good, thanks," Karen answered.

"Are things really okay?" I asked sympathetically. "I heard you were going through a divorce."

"Yeah, I'm getting a divorce," Karen answered. "It takes time."

"Okay," I nodded. "I'm sure that's painful. Please let me know if there is anything I can do to help you in this transition."

I gave her grace, figuring the transition caused her to seek male attention.

But things escalated. Karen and a male NCO pranked one another. She'd put a dead bird in his car, and he'd put dead bugs in hers. Yeah. I'm supervising a second grader. All day long, the two of them kept it up, which meant they weren't working. I was responsible for her; her male friend reported to a different Senior Non-Commissioned Officer (SNCO).

Then they walked together from the back of the warehouse one day—while she fixed her hair and tucked in her shirt, as he adjusted his belt. It wasn't long before everyone in the warehouse was talking about the two of them.

I was pulled to the side by the NCO's supervisor, who let me know that she had already verbally counseled him on unprofessional relationships and perception. Due to this being his second infraction, the first of which resulted in him leaving his last base (infidelity to his wife), he needed to tread carefully. Once she let me know that she had handled her NCO, I needed to have a talk with Karen—and sooner rather than later.

"I want you to have fun and relax, but we have to make sure we are doing the mission, Karen," I started. "If you're constantly running up and down the stairs chasing each other and bump into the flight commander, that would be a bad day for all of us. I want to make sure we're maintaining a professional work environment. You are new here, and people are focused on you guys. I'm not saying people are watching you, but they want to make sure you all are adjusting well and getting what you all need. Since you are new and going through a divorce, people are looking out for you—not in a bad way, but they want to make sure you have the resources you need. And I want to know if there is anything I can do."

She sat quietly.

"This is the fourth time that someone has spoken to you about perception and unprofessional behavior," I continued. "Once with the superintendent, once from our boss, another time from someone outside our squadron, and now me. I'm not accusing you of anything, but I want to make sure you are practicing professionalism and building professional relationships while you're here. And it's imperative that you know what we can and cannot do while on duty. Again, I am not accusing you of anything. To make sure you know what's expected of you, I printed out a regulation, as I do with all material I reference, so we can look at each description."

I walked her through regulations on professional relationships, professionalism at the workplace, and AFI36-2909 2.4: "Relationship of unprofessional conduct to other provisions of the Uniform Code of Military Justice."

I wanted to be sure she understood what the Air Force deemed unprofessional—and the repercussions if she decided to ignore regulations. I wanted to help her correct her behavior or, at the very least, figure out what the other three NCOs who spoke to her saw when they verbally counselled her.

I had another reason, too. My mom taught me at an early age that if something is worth doing, do it right the first time. I didn't want to have this same conversation over and over again on the same issue. Instead, I hoped she would learn from it, and the two of us could move on to a productive, positive working relationship.

"Are all E5s such assholes?" she asked me, taking me aback a little. I might have said things like that in my head, but I always tried to speak respectfully to my superiors, even when it took every bit of restraint I could muster.

"I'm not accusing you of anything," I clarified in case she missed my message. "I just want you to take care of your reputation and not come across as a goof-off who doesn't care about the mission. Don't give anyone a reason to come down on you."

In hindsight, it didn't matter what I said or intended to say. She got it in her head (wrongfully) that I had just accused her of adultery, and she went off the rails. She spent less time messing around, which was good. But instead of spending more time working, she used that extra time to trash me, calling me a bitch to anyone who would give her a second.

Which turned out to be a lot of people, including my boss. As a result, I needed to explain to my boss my entire conversation and offer the airman my apology.

"To be clear, I never said that you were having an affair," I started. "If that's what you took away from our conversation, I apologize. That was not my intent. My intent was to let you know that perception is reality. That's all. Don't give people a reason to say anything negative about you."

She might have taken this feedback better had the NCO in question not received such a stiff reprimand for his conduct from his own superiors. The SNCO had no choice, since this was the third time he'd been talked to. After this final discussion, the NCO stopped cold his interactions with Karen. He wouldn't even look her way.

Of course, she blamed me for her failed "friendship." She took it all out on me. "Now he won't talk to me. Basically, you will not allow me to have any friends."

When I'd been on her side of the table, my superiors rarely let me talk. I decided to let her vent.

"If you want to blame me, that's fine," I said as I leaned back and listened.

"I swear, Sgt. Clark," Karen continued, "if you ever talk to me again or even look at me, I promise I will report you to the inspector general! I'm going to have to pray hard for you tonight, because where I'm from, I would have been throwing punches already. The stuff you did, people in New York would have already punched you in the mouth."

It's like I had DOORMAT printed on my forehead, because she placed her anger on me, yet she still had enough energy left over to complain about me to everyone within earshot. Venting can do two things. One, it can help the anger dissipate. That had been my hope. Or two, the negativity can spread. Our squadron already had low morale, and I'd learned the hard way that every team needs a scapegoat—someone to blame for everything, ranging from severed relationships to no ice cream in the canteen.

The more people took her rants to heart, the more I hated my job. I got it from all sides. However, a few of my peers encouraged me to stay level-headed.

"She's grieving," one reminded me. "Her marriage is in the crapper, and she needs someone to blame. If you give her paperwork right now, you'll be the one with your head on the chopping block. It's like kicking her when she's down."

"I don't know about you," I replied. "But they certainly don't pay me enough to deal with this."

Yet I understood what my colleague meant. If I played the role of the active bad guy and came down hard on her, I would be committing career suicide. It's one thing to go to your boss about something as cut-and-dried as policies, regs, and rules. It's another thing to whine, "My employee is being mean to me."

So I shut my mouth. Not because I wanted to. My supervisor had advised me to take it and let her lash out on me. After all, she reasoned, in Karen's mind, I'd just accused her of adultery. Besides, I still had the stench of the Kramer situation hovering around me. And since SJ had received a no-contact order, I fell back into the shithole work cycle.

In September, another crisis in the squadron took up any time I would have otherwise spent worrying about my renegade Karen. Our flight commander issued a mass punishment to all NCOs in my flight due to failing our training program. Our failure involved not keeping training records up to date. All NCOs keep training records of their own airmen, and ours were not correct.

For punishment, we were given twelve-hour shifts, working through the weekend to get all our records correct. Since we were given group punishment, we had no way of knowing who got it right or wrong. I suppose the goal was to let us figure it out ourselves; however, we were shooting in the dark.

I later learned my records were right. I didn't need to change them. Had we known this at the beginning, my fellow-NCOs could have changed their records to resemble mine. Instead, we all worked frantically to guess correctly. As a result, our resentment and frustration levels spiked.

Even before this mass punishment, no one liked our squadron. We were considered the fuck-ups of the base. Our morale hit a new low.

"This shit sucks so bad I'm going to start drinking again," one coworker groused. He smelled like booze. Either he was hungover from the night before, or he may have started drinking before coming to work.

A new NCO to our squadron came to me crying at work.

"Is it always this bad?" she asked. "Because I don't think I can take it much longer."

I tried to console her, which basically meant I didn't tell her that I, too, felt like I was going to have a nervous breakdown.

Back in Virginia, I'd watched a coworker become unhinged, talking about shooting up the place and bombing the command headquarters, so I took these comments and tears seriously.

Around the same day, I escorted an airman to the chaplain's office because he felt suicidal, so I did a warm hand-off to a chaplain who could assist him.

That same day, another NCO said, "I'm so fed up. Fuck this, and fuck everybody. I'm about to blow this place up." He then picked up a 4"x4" piece of lumber and smashed a piece of equipment with it while the rest of us stood watching.

Once he calmed down, I pulled him aside.

"Look, we are all tired of this shit," I agreed. "But you can't lose your cool, especially in front of junior airmen. They look up to us as their leaders. If you're pissed off, you gotta go somewhere else to let it out. I'll even cover for you. But don't do it here in public."

That same week an airman was teased so relentlessly that his mental health went into a tailspin. While working in the warehouse, a pipe slipped and sliced open his finger. Another airman took him to the base clinic. While the medic there tried to numb his finger to sew it up, the airman groaned in agony. The airman who escorted him secretly recorded his moans of pain, then sent it around to everyone in the unit. Instead of the people around him offering support, they played the video and mocked him mercilessly for making crying noises.

I took my concerns to my leadership. My flight commander, the one who ordered the punishment, didn't want to hear it. He directed all complaints, questions, and concerns to his section chief, who was also my supervisor.

"They brought it on themselves," the section chief shrugged, as she told our NCO group, "They're fine. And let me just remind you of something. What happens here, stays here. I don't want you complaining about this to anyone outside of my house. Do you hear me? Now get back to work."

I couldn't live with myself if someone took their life or hurt others while I saw it coming yet did nothing. In August, I wrote an anonymous letter to my chief outlining my concerns.

Unlike my flight leadership, the chief got involved based on my anonymous letter, and he held one-on-ones with everyone in the squadron. When he finished talking with everyone, he ordered the weekend hours to stop.

While I was grateful for that, my flight commander became irate that "someone snitched" and "made me look bad." But I didn't care. He could be mad at me if he chose to. I knew that I had done the right thing.

Are you familiar with the adage, *No good deed goes unpunished?* Yet again in my military career, I would experience this truth.

As this situation resolved itself, my renegade airman had been making the rounds to get me removed from the section.

"I can't work with her any longer," Karen cried to my commander. "I'm afraid of her."

My flight commander, believing that I had written the anonymous letter to the chief, took action. He removed my direct reports and moved me out of that section. That's like taking a brush away from a painter. What good is a supervisor with no one to supervise?

But he wasn't finished. He stepped in to try to resolve the situation between me and Karen. He called a meeting with me, Karen, my supervisor, Karen's supervisor, the superintendent, the assistant flight commander, and my section chief. Let's just say the outcome went over like a fart in church.

Immediately after that failed attempt to resolve the issue, my flight commander conducted another meeting, as Karen continued to threaten me with involving the inspector general if my leadership didn't take away my sergeant stripe! The attendance at this meeting made the earlier one feel small. The flight commander ordered eleven of us to a room as a final attempt to get a resolution.

At the beginning, I felt like the chief stood with me.

"Can you see that Sergeant Clark meant well? She was trying to mentor and guide you, to keep you out of trouble. She had good intentions. I don't see what the problem is."

That's when Karen interrupted him. "She absolutely ruined my life!" Karen went on to lie about my conversations with her, adding that I had been "gunning" for her since she first arrived. And she suggested that I had created a hostile workplace by threatening her and her Air Force career.

When she finished, others piled on with their own issues with me. Eventually, the gloves came off. Throughout the meeting, I was called a *bitch, liar, crazy,* and *stupid.* They took turns laughing at me and adding insults like, "There's something wrong with her." And when I say *they,* I mean from the highest officer to the lowest airman, they took jabs at me and my character.

"Can you make this stop?" I asked the chief. "Or at least rein it in a bit?"

"No, this is good," he said. "I think it's good to get it all out."

Good for whom? I wondered.

175

"They are berating me, and you're allowing this to happen," I shook my head.

"Yeah, I am," the chief said, looking at me. "Because you wrote that anonymous letter."

Boom! Everyone in the room gasped.

"She wrote that letter?" whispers echoed around the room.

"It was supposed to be anonymous, but thanks for that," I said to the chief, letting sarcasm creep into my words.

"Yes," the chief confirmed to everyone present. "She wrote the letter. And," he let the silence hang in the air for a moment, "I've learned that Sergeant Clark was apparently assaulted at her last base, too."

"Oh my God," my flight chief shouted. "That's rich!"

A new round of "liar" comments floated across the room while my face burned with anger, hurt, and helplessness.

The chief chuckled with them for a minute before concluding the meeting by adding, "Well, Clark, it looks like you made quite a few enemies here today. I want you to take a long look in the mirror. Ask yourself what you've done to these people to make them feel this way towards you. I don't have a single person in my entire life who doesn't like me. Do you know how many enemies you should have? You should have *none*. So if even just one person doesn't like you, you need to reevaluate yourself."

After the meeting ended, I asked the chief if I could talk to him one-on-one.

"No, I don't think so," he answered. "We are done talking for today. Let your supervisor know if you still want to talk later, and she'll schedule you in."

I felt like I'd been in a cage match with a group of savages. For an hour, I got pummeled from my former airman, my supervisor, and the chief. Once again, "seeing something, saying something" had come back on me. The issue moved from effective, appropriate supervision to my *audacity* in taking inhouse problems up the chain of command. Even though everyone in the squadron felt relief when our punishment was suspended, the message that came out of this room was one I knew too well: *Kia is a nark*. And if you've ever watched a crime show on television, you know how kindly people take to narks.

When I got home, I slipped into my closet, closed the door, sunk to the floor, and cried for an hour. Once I regained my composure, I changed my clothes and checked my phone. My superintendent texted me later.

> "Hey S.Sgt. [Clark]. Just checking on ya to see how your evening is going? Rough day today. Tomorrow is a new day with new beginnings! I do not know how your last base was mentioned. I know Chief met

with a lot of people, but I do not know. I was kinda torn this evening on calling to check on you, hoping you got some positive insight in today's mtg, but also concerned you lost trust in me. But I'm proud that you held your composure and bearing! You should be proud of that!"

This did nothing to console or comfort me. Coming from a person who literally sat in the corner in the room while a pack of wolves attacked me while she said NOTHING on my behalf, I could find no positive insights or peace. Her pride in me burned under my skin like a smile offered from the front row at my execution.

I texted her back, "I feel sick and depressed, like this might break me. My chest is hurting, I cannot do this anymore."

She replied, "Treat yourself to a mani-pedi in the morning. I'll let your supervisor know that you won't be in tomorrow. Enjoy your day off."

I didn't need a mani-pedi. I didn't need a day off as a consolation prize for my shredded dignity and maligned character. I needed someone to care, to listen to me, and to take me as I was—instead of buying into the negative reputation I'd earned while trying to navigate *the right thing* and *the Air Force thing*. I needed help for my mental health. I felt at the end. My superintendent didn't offer to come see me, give me a suicide hotline number, or anything relevant to my shattered emotions. Instead, she offered me glossy nails.

I passed on the mani-pedi. Within a couple of days, I did get some "time off." I was admitted to the hospital after I started vomiting blood.

"I'm not surprised," my supervisor told me. "It's probably because of all the stress."

Compassion at its finest.

Later that week, I learned that my elite assignment with the Joint Communication Support Element had been cancelled. This top-secret project would have had me working with the Army, and its appointment to this group was as prestigious as it gets for an E5. Some assignments were better, but this one offered a joint environment position where I could learn and contribute.

I learned my assignment had been pulled from me when I got called into a meeting with my squadron commander, chief, and supervisor. My commander informed me that my assignment was cancelled on the advice from my chief.

My chief had already humiliated me in from of my entire leadership team, but he wasn't done. He then heaped additional punishment and unfavorable action on me by removing this work assignment, all as a result of the *protected* communication I sent him.

"May I ask why my assignment was canceled, sir?" I asked my chief when I learned he had pulled it from under me.

"Because you wrote that letter," he said without looking up.

"Because I spoke up to take care of the airmen in my squadron when they wanted to kill themselves or blow up the base?" I asked, trying to keep sarcasm out of my voice.

"No," he replied coldly. "This is for not knowing when to leave things alone. You're dismissed."

Thanksgiving is coming, Kia, I told myself. *If you can hold on until then, you can get out of here for a few days and put this behind you. Hold on until Thanksgiving.*

Before leaving for Thanksgiving, I went to the inspector general, to inform them that my leadership canceled my assignment based on "protected communication."

The person I met with at the IG office was kind and listened carefully, which gave me hopes that the IG would intervene on my behalf.

As soon as he started talking, I learned that I was doomed to lose.

"Your chief, obviously, is at the top of your chain of command," he said, raising his hand to give me a graphic indication of where the chief stood in the scheme of things. "You are down here," he continued, lowering his hand down a couple of feet. "If you're going up against him, you better have a spotless record. Because I can tell you they will look for any dirt on you. I highly advise that you sleep on this decision and really think if you're ready to take on a chief, because I know about the case with S.Sgt. Kramer. I think it's best you really think on this."

I didn't respond immediately, so he continued.

"You say you wrote a note to the chief but didn't keep a copy. If the chief says he never received a letter, and you don't have proof. Then what?"

I conceded. Then he proceeded to help me with the withdrawal request form to stop my complaint from moving forward.

My conversation with the IG took me back to when I reported the low morale and suicidal comments made on my team. What struck me again, as if for the first time, was that my leadership took great lengths to quell the complaining and identify the rat, but not even one of them took interest in learning more about the airman who talked about ending his life. They were so obsessed with damage control and image management, they failed in the more critical role of leadership: compassion.

Chapter 24

A Time of Thanksgiving

"Every new beginning comes from some other beginning's end."

~ Seneca

I N THE DAYS leading up to Thanksgiving, my mom called me at work. She had been working on base at the hotel, and she had concerns about a coworker of hers who worked in the coffee shop.

"I don't know what to do," my mom said in a panic. "She said she's going to kill her boss, and then kill herself. Her boss isn't taking her seriously or letting anyone help her. We don't know what to do. Can you help?" she pleaded.

I dropped everything and rushed to the hotel on base.

"Where is she?" I asked when I saw my mom.

Mom pointed to the coffee shop around the corner, and I hurried inside.

"Hey, girl," I said as I walked in. "What's going on?"

Amanda leaned against the counter, tears dropping from her face.

She told me that she couldn't take it anymore, and she planned to kill her boss and then herself.

"Okay," I said soothingly as I put my arm around her shoulder. "Do you know what we're going to do? We're going to take you to the base clinic down the road. They will be able to help you, okay?"

"I can't go anywhere," Amanda wailed. "The only way my boss will let me out of here is if she is dead first. I'm gonna start stabbing or shooting people here!"

"No, honey, no," I continued to try to keep her calm. "No one needs to get killed. When you start thinking about hurting yourself or others, that becomes more important than who's going to serve the coffee to customers. Let me go talk to your boss, okay?"

I pulled her boss aside and explained Amanda's mental state, explaining that she was talking about murder and suicide.

"Well, who's going to cover her shift?" her boss demanded. "She pulls this shit all the time. She's always suicidal. What I want to know is *who's going to replace her?*"

"Well, maybe you can cover her shift for a while," I suggested. "She is suicidal. But before she takes her own life, she talked about killing you. So maybe the best thing to do is to get her the help she needs right now. And we can figure the rest out later."

Her boss scowled and said nothing.

"We have to go," I pressed on. "You're okay with me taking her to the clinic?" "Fine," she said to me. Then looking at Amanda, she added, "But you ain't getting paid for this!"

Driving to the clinic, I realized that I had no idea if Amanda had a gun or knife on her. It never occurred to me that I might be in any danger. But that started to creep into my mind, so I tried continuously to keep her calm.

"I know you're having a hard time right now, Amanda," I consoled her. "We all get like that. And I'm going to stay with you until you're feeling better, okay?"

Before leaving the car, I checked in with my new boss, M.Sgt. Dunn, so she'd know why I wasn't at my station.

"I just left the hotel with a suicidal woman. At the clinic now," I texted her.

"Why were you at the hotel?" M. Sgt. Dunn asked.

"Mom called. Her coworker falling apart," I responded. "No one here knew what to do, so my mom called me for help."

"Why didn't you let me know?" she asked.

"I informed coworker," I texted back, "but I wanted to let you know as well."

"Expect to be disciplined on return," M. Sgt. Dunn replied. "This is why we have security forces. You put yourself in unnecessary risk. Stay at clinic. I'm coming for you."

I acknowledged her message and took Amanda inside.

I didn't know the clinic was a cellular dead zone. I learned later that M.Sgt. Dunn called and texted me repeatedly, but I didn't get any of her messages. Unable to find me, she went to the hotel and talked with Amanda's supervisor about what happened. I also learned Dunn and Amanda's boss had some words, and both ended up angry about how the situation got handled.

But none of that mattered now. What did matter was that Amanda could get help. I asked her to sit down while I checked in at the front desk.

"This girl behind me?" I said softly to woman at the front desk. "She needs help. She's actively homicidal and suicidal."

"And she's military?" the woman asked.

"No, she's a civilian," I responded. "But she works here on the base."

"Sorry," the woman shook her head. "We can't serve civilians."

I went to another part of the clinic and had the same conversation with another person working the front desk. And I got the same answer.

Finally, I saw a sign on the wall for the mental health clinic, and I took Amanda upstairs to have her sit in the waiting room while I talked to the person behind the desk.

After explaining the situation *again* and getting the same response *again*, I pushed back.

"You're a mental health clinic," I said, trying not to raise my voice. "You are supposed to help people regardless of their military status. She's hurting, and she's talking about hurting herself and others."

"Sorry, there's nothing we can do," she answered.

Amanda heard this entire conversation, and she sobbed as she heard yet another refusal by medical staff to help her in her broken state.

"Come on, honey," I said to Amanda. "We're not leaving until you get help. It's going to be okay."

I returned downstairs and walked into Flight Medicine, a specialty clinic for pilots. Desperate, I took a more direct, stern approach at the front desk after I got Amanda seated.

"This woman behind me is suicidal and homicidal," I said. "I know this clinic is for military personal. She's a civilian. I know your area of expertise involves flyers, and she's not a pilot. But I'm not leaving here until she is seen by a doctor. I don't care who you have back there or who you need to approve this. I will not move from this spot until a doctor comes out and takes her into a treatment room. Do you understand?"

Instead of an argument or a brush-off, the person behind the desk nodded. Then she went into the back and returned with a doctor.

"Hi, sergeant," a doctor said as he came out from the back. "I'd be happy to see you and your friend now."

He led us in the back and started talking with Amanda. At the same time, the desk clerk alerted security forces. Eventually, civilian police arrived and joined us in the doctor's office.

Amanda asked me to stay with her while she talked with the doctor. She explained all the things she'd been thinking and feeling to the doctor. She talked about how disrespectful her boss was to her, and she also mentioned the stress of her and her husband working all the time and never seeing their children. All these factors weighed heavily on her, causing her to lose hope and want to end her life.

When I returned to work, my flight commander and superintendent brought me into a room and asked me what happened. I explained the whole story.

"You did the right thing as far as I can see," my flight commander told me. "Thank you for helping her. Stand by if anyone needs more information from you."

Then I got an email from the medical group commander.

"S.Sgt. Clark," the email began, "Thank you so much for your persistence in not giving up when you ran into resistance at the clinic yesterday. This situation and your actions identified processes within my clinic that need to be fixed immediately. I will be conducting additional training with staff. NO ONE SHOULD EVER BE TURNED AWAY WHEN THEY NEED IMMEDIATE HELP, ESPECIALLY THOSE WITH SUICIDAL OR HOMICIDAL THOUGHTS. Thank you again for your courage for pushing until she got help. You may have saved more than one life with your actions."

The med group commander cc'd several of my leaders on his email, and I got praise and thanks from across the base. Even M.Sgt. Dunn reached out.

"I'm so proud of you," she wrote. "Please make sure they know that I was helping you as well."

Helping me? I thought to myself. *Telling me to expect punishment when I returned was help?*

Too many of my Air Force leaders are experts in deflecting blame, throwing their airmen under the bus when things go sideways; but those same people strain their neck muscles trying to photobomb any situations where praise is involved. I know that this is a human tendency, not one that's unique to the Air Force. Accountability should be a given, and leaders should be as quick to raise their hands when blame is assigned as when credit is earned.

Eventually, Amanda went to a facility for further evaluation and the help she needed. A week later, she returned to work feeling much better. And she asked my mom if she would ask me to visit her.

"Thank you so much for everything you did for me," Amanda said as she rushed over to give me a bear hug. "I got so confused and scared, I didn't know what to do. Thanks to you, I got help. You are an angel."

Heading into the Thanksgiving holiday, I had three things to be most thankful for. One, I had saved a woman's life. I didn't care about the praise or recognition, although it was nice. I felt good that I had the chance to help another human being. I didn't always feel like I made a difference in my military service, but I was grateful that I had been in the right place at the right time to make a difference in the life of one person. Two, since my dad invited us to his place, I would get to see him. This would be the second time he would see Reign since she was born over a year before. And three, I rejoiced at any opportunity to get off the base. They would have a hard time blaming me for anything that happened when I wasn't there.

Little did I know, I would be given one more reason to be grateful on that trip. Driving to Tennessee for the holiday, SJ stopped for a fill-up while I stepped inside to buy a bag of Twizzlers. You know me and food. I always need something within reach. I set the bag near the gear shift while I munched on one.

Once SJ finished pumping gas, he got in the car, buckled up, and put the car in reverse. In the process, he slammed the gear shift into the bag of candy, sending my Twizzlers flying through the car.

I laughed as the Twizzlers finished bouncing off the ceiling and raining down on us.

"This wouldn't have fucking happened if you didn't put them in my way," he screamed. "This is your fucking fault!"

You just shredded my damn Twizzlers in the gear shift, but you don't see me getting mad! Shit happens. If I can laugh, you should be able to laugh, too, I thought but didn't dare say.

Before I could say anything, he picked up pieces of the candy and started throwing them at me in a rage.

"I'm sorry," I said, trying to shield my eyes. "I wasn't laughing at you!"

"Your fucking fault!" he continued to say.

He wasn't done blowing up. He took a Twizzler and started to whip me with it while Reign screamed from the backseat.

Have you ever been beaten with a damn Twizzler? That shit hurt. I mean, it left welts on my arm and face. Even though we had not yet gotten to Tennessee, I knew at that moment I would not stay with SJ. The thought of him choking me to death scared me less than staying with him one more day. Finally, I had reached the end, but I kept that to myself. I savored the time with my dad, remembering with gratitude my dad's warning about *rushing into marrying SJ* and wishing to God I had listened. On the drive home, my mind kicked into action, planning my next steps.

The day after we returned from our Thanksgiving holiday, I went online to start filing for divorce. I thought I would feel angry as I filled in each line, but I didn't. I had no feelings left. My emotions over him had died. SJ surprised me when he didn't go into a rage, threaten me, or try to kill me when I told him my plan to divorce him.

Florida doesn't recognize legal separation, but I needed a sense of finality. I wrote a letter stating that as of November 28, 2017, SJ and I were separated, awaiting the divorce process. After getting the letter notarized, I signed it and gave SJ a copy. Emotionally, I was gone. With this letter, I burned my bridges. I could not and would not turn back.

Do-overs. Starting shortly after my first date with SJ, I wished I had a few do-overs. Even one would have changed the bed I'd made for myself. When he called me a whore, I should have left him. After the first time he raged at me, I should have been done with him. When he proposed, I should have said *no*. I should have jumped out of the moving car while driving to the courthouse. I should have, I should have, I *should have*. For years, I beat myself up over what I should have done, longing for a do-over. But still I stayed, letting fear take hold of my heart.

But this time was different. I wasn't dreaming about a do-over. I was taking back control of my life, one decision at a time.

When I handed SJ his copy of the document, I told him, "I am done with you." I'd finally learned that we aren't given do-overs. We must fight for them.

That Thanksgiving, I had more reasons to be grateful than I'd had in years. I finally arrived at the beginning of the end. No longer would I put SJ over my family. No longer would I dwell in a home where I felt unsafe and where my precious infant daughter heard yelling and saw his rage. I was done having secret, nervous conversations with my few confidantes about how miserable things were at home. That letter was more than my first real step forward. It was also my way of dropkicking this door closed.

We were both scheduled to deploy, me in January and SJ in March. Since we would both be out of the country, we decided to rent a small house and ask my mother to move in to watch Reign while we were gone. That meant the two of us would be under the same roof for a month before I left and another month when SJ returned, and our lease expired in December.

I wanted to file and wrap up the divorce before leaving the country, because I knew from experience how manipulative and cruel SJ could be when he went into a rage. Unfortunately, things with SJ never happened in a straight line or quickly.

Before I left the country, SJ's first stall tactic involved him transforming himself back into a medieval knight, a combination of SJ the Fabulous and SJ the Warm.

"I love you so much," he told me with his head hanging down to his chest. "I'm so sorry. I know I haven't been good. I'll do anything to keep you."

Blah blah blah. Keep talking, asshole. I'll be gone in a matter of days. I reminded myself I would have enough fuel in my escape pod by the time I returned from deployment.

Then he tried the old "I tried so hard to be the best man, husband, and father I could be for you and our daughter. Everyone always leaves me. My ex-wife left me, and now you're wanting to leave me. I'm just this unlovable man who is clearly unworthy of love."

"I pray to God that he gives me a second chance to prove to you how much I love you," he cried, trying on his SJ the Woeful suit of armor. "But if the Lord doesn't will it, then I wish you the best. Please keep your wedding ring just in case you come back to me. That ring will always be yours," he offered magnanimously.

Sure, babe. I'll hold on to that ring. Until I pawn it, I thought.

After thinking through things, I decided not to file for divorce until our deployments were over. But make no mistake about it, I didn't have a change of heart. Rather, I made this choice pragmatically. Had I filed and either of us received moving orders, we would have needed to scramble to figure out custody for Reign. I'm glad I waited.

Just before I left the country in January of 2018, I was part of an all-call meeting where I was awarded at a coining ceremony. A coining ceremony is a special recognition for airmen who do something well above the call of duty. My wing commander coined me for my efforts to help Amanda get the mental health services she needed.

Throughout each year, the Air Force conducts Green Dot training, which includes sexual assault awareness and suicide prevention. In all military branches, leadership is keen to show off the efficacy of their training efforts. Since I'd been recognized by medical doctors higher up the ranks, my leadership couldn't ignore my actions. My supervisor let me know that the coin I received could create multiple wins. First, it acknowledged my action, making me out to be a hero. Second, if I mentioned that Green Dot training was the reason I took the actions I did, it would make my leaders and the training program look good. Finally, it might help me get back into the good graces of my leaders.

I didn't help Amanda because of Green Dot training. *Do I really need a training program to tell me what's right or to help someone who's struggling? No. But whatever.* If they wanted me to be the poster child of Green Dot training, I happily accepted.

Shortly after the coining ceremony, I boarded a plane for my second deployment, this time to an undisclosed location (one that I can't divulge). But unlike my first deployment, this one encapsulated the highs and lows of my Air Force experiences in a six-month period.

Chapter 25

Undisclosed Location

"It was the best of times, it was the worst of times…"

~ Charles Dickens

F OR MANY AIRMEN, deployed life serves as an escape from the doldrums
of home. Think of it as a social time on foreign soil where you can meet new
people and enjoy new experiences.

T.Sgt. Chipo, my immediate supervisor on deployment, picked me up at my tent
the morning after my arrival.

"Do you want to join us for wing night?" he asked as we headed to breakfast. He
explained that every Wednesday night, they would take a bus full of airmen to another
base an hour away for a taste of America: Buffalo wings.

Excited to meet new people, I boarded the bus three days after I arrived at my
new station. M.Sgt. Barren (Bear for short) sat beside me at the front of the bus. Since
I sat up front, I had no idea of all of the free space behind me where Bear could have
sat, so I made room for him to sit down.

"Hey, I'm M.Sgt. Barren," he said, extending his hand for me to shake. "Friends
call me Bear. Let me ask you something. What are your goals for this deployment?"

"Well, I want to learn my job and add value while I'm here," I answered, not
knowing what else to say.

"Great work goal," he nodded. "But I'm talking personal goals. For example, one
of my goals is to start a book club. I'm going to leave this deployment smarter than
when I entered it, and part of that means I want to read at least fifteen books before
I go back home."

"That's great," I said.

"Hey, have you watched *Game of Thrones?*" he asked.

"No, I haven't seen it," I answered.

"Well, you should," he said excitedly. "It's awesome. And fortunately, you have six months to do it. It's not like there are a lot of other things to do here."

While Bear and I talked at the front of the bus during the hour-long drive, I could hear T.Sgt. Chipo talking and laughing with others in the back of the bus.

After eating, we piled back on the bus, and I sat in the same seat at the front. Bear slipped in next to me again. We talked non-stop the entire time. I was grateful to meet someone who was easy to talk to.

As we pulled up to the base, we got off the bus, and Bear stood next to me.

"I had such a wonderful time getting to know you," he said. "I can't believe I'm saying this, but I almost wish the drive were longer!"

"Me, too!" I said sincerely. "Talking with you made the time fly by."

"Hey," he continued. "Do you mind if I give you a hug?"

"I'm not much of a hugger," I said and raised my hands in front of me in the universal stop gesture.

"Ha!" he said. "You're so funny!" Then he grabbed my shoulder and gave me a side-hug.

"You're the best," he said, and then he slipped his hand under my arm and squeezed my breast.

"What the hell!" I reacted, pushing him away.

Bear simply laughed and walked away.

I was livid. I found T.Sgt. Chipo immediately. "This dude," I said pointing to Bear while he walked away. "He just felt me up."

"Hahahaha!" he guffawed. "There goes Bear again, reeling in another one! He got you good." He and my coworker laughed together as if this were the funniest thing they'd ever heard.

T.Sgt. Chipo would rotate out in three months. For the rest of his time there, he kept a tally mark on the white board of how many men groped me or made an overt pass at me. Mind you, those were only the ones he witnessed. Bear became the first tick mark.

"Clark's got another admirer," he'd say each time he'd observe someone crossing the line with me. Then he'd make another tick on the board. By the time he left, the count stood at twelve.

Very quickly after Bear groped me, I learned that I was the unwilling part of a game for some of the men. They'd flirt with me, grope me, or try to sleep with me.

So just as quickly, I took on the persona of the angry Black female. I stood my ground and didn't allow others to objectify or demean me as a woman in the military.

Soon, I took charge of my own section. Eventually, another NCO, S.Sgt. Campbell, joined the team reporting to me. Our area had six jobs with me owning responsibility for three of them, and another supervisor for the other three. Since we were a small unit with relatively few people and many tasks, the work was stringent. I needed to be taken seriously. In no time, I had the respect of those working under my areas. I had worked hard to demonstrate to others I was smart, competent, and determined. In other words, I refused to be reduced to *a pair of boobs and an ass.*

I did my job very well. When I left deployment, I received a decoration. It had been that any airmen on deployment would get a decoration, but they had become very meticulous by this point with what merited special awards. I worked hard for mine.

One day in the warehouse, a technical sergeant asked for a pair of socks. No problem. I issued him a new pair. I wouldn't have remembered it had it ended there.

Later that day, I got an IM on the government computer.

"Thanks so much for the socks. You're a rock star," he wrote.

Yeah, that's me. What Hendrix did with guitars, I do with socks, I thought to myself. "Okay, thanks," I replied. "Is there something else you need?"

"Just wondering," he replied. "Do you watch *Game of Thrones?*"

What the hell is with Game of Thrones? I wondered. *Is this a new pick-up line? Do I need to look it up on Urban Dictionary?*

"No," I answered, giving no encouragement. "But is there anything else that you need?"

"Are you married? I don't care either way, just curious," he wrote. "Do you want to come to my room, watch some TV, and chill? Just so you know, there's no male/female threshold rule, so you can come over to my room anytime."

"I'm married to the Lord. And the only chilling that I want to do is in the Kingdom of Heaven," I typed, slamming the door shut.

"All you had to do was say no, airman," came his reply.

In the Air Force, calling someone an airman is the same as saying bitch.

I had learned long ago that I could tolerate people making a pass at me, or I could stand firm against the nonsense and be called a bitch. I chose to be the bitch. If that's how others wanted to view me, fine. I had a job to do and the determination to do it better than anyone who worked it before me.

In addition to calling me "airman," he reported me to T.Sgt. Chipo, claiming I *didn't know how to provide proper customer service.* He said I should be removed from my position and replaced by someone else. He also reported that I was *rude*, adding that if I weren't removed, he'd go to the chief. While I wasn't removed, T.Sgt. Chipo spoke to me about being more friendly and approachable.

"Stop being so uptight. If someone takes a little interest in you, see it as a compliment, okay? You don't have to go all feminazi on his ass! Or I'm going to have to remove you from your position," Chipo added.

S.Sgt. Hernandez worked on the other side of my warehouse, and he also took an interest in me—hard. I stayed in the tents on the east side of the base, and he stayed in the dorms on the west side. It's about a ten-minute walk from one side to the other.

Whenever I'd come outside my tent, I'd see him in my area on the opposite side of camp. Since the dining area, gym, and recreation hall were in the middle of the compound, I could think of no reason why he'd be hanging out around my tent. I mean, the only thing you would do in your tent is sleep. There was nothing for him over here.

Then I started seeing him pacing back and forth in front of the window in the wall that separated our work areas while I worked.

"Sergeant Clark," one of my airmen asked. "Why is he always here?"

"I don't know, and I don't care," I answered. "If he's got no work to do, his days must be very long and boring. Let's just get back to work."

I didn't want my airmen seeing this stalker. I adopted an *out of sight, out of mind* approach to his behavior. *If I don't acknowledge him, he's not here.*

That worked until he walked up and opened his mouth.

"What tent do you stay in?" he asked out of nowhere.

"Go back to your job, sergeant," I responded.

"Come on," he persisted. "When are you going to be in that tent? I like the tent, and I want to come over there to see you."

"Get back to your work," I said again.

"Hey, sergeant," one of my airmen said. "I don't think she's interested. Do you want to go back to your work area now, sir?"

I loved my airmen! This sweet, nerdy little guy stood up for me. And he even managed to say "sir" and show undeserved respect in the process.

After that, I asked my airmen to walk me back to my tent after our shift ended.

"You might not know it by looking around," I told them, "but chivalry is not dead. If you expect to attract the attention of a lady, let me tell you how to show that you're a gentleman: open doors for her, pull out her chair at the table, don't honk your horn for her to run out of her house when you pick her up for a date, and don't send her off in a bad neighborhood alone. If you want to practice, you can walk me back to my tent after work."

Under the guise of teaching them chivalry, I got escorted back to my tent each night after work. They knew me well enough to know that I wasn't wanting them to hit on me or see me as the lady they might be able to attract. We worked so closely together and became friends, so our walks at the end of shifts became our time to talk about what went well during the day. Our walk and talk time also kept me from having to walk through my own bad neighborhood.

But that didn't help in the morning.

One day I woke at about 4 a.m. to use the restroom, and S.Sgt. Hernandez stood outside my tent.

"What the fuck are you doing here?" I demanded.

"Just out for a morning walk," he said coldly. "But I found your tent."

What a lunatic!

When I came out of the restroom, S.Sgt. Hernandez made a show of checking his watch.

"You were in there for four minutes and thirty-eight seconds," he smiled.

Again, what the piss? I thought as I did an about-face and walked back in my tent.

Once I got there, I texted my coworker, S.Sgt. Campbell, and asked if he could come get me to have breakfast together.

Pathetic. During the first month, I asked my airmen to walk me home to my tent. Now into month three, I'm asking a coworker to get me in the morning. All because of a stalker.

The next time S.Sgt. Hernandez appeared, he came to my desk.

"Can I get my workout gear?" he asked with a friendly smile.

"I'll have someone come help you," I answered.

"Do you really want to turn me away?" he asked. "What's your boss going to say when you refuse to serve a customer by simply getting him his gear? I could talk to Sgt. Chipo and maybe the chief if you'd like to be removed from your position."

I asked T.Sgt. Chipo to intervene.

"Clark, just get him his gear," T.Sgt. Chipo said in an exasperated way. "It's not a big deal."

Returning to the desk, I saw S.Sgt. Hernandez's gear was stored in a container at the top of a large shelf. "Go ahead," S.Sgt. Hernandez said with an evil smile. "I'll wait. You can get it. Just use the ladder."

"I'll get it and bring it to you once I'm ready to issue it out," I said coldly.

"No, ma'am," he said with a sneer. "I'll wait right here."

While I climbed the ladder and reached overhead to get his gear, he kept talking from the ground.

"Damn, that looks really fine from down here," he said.

I grabbed his stuff as quickly as I could and handed it to him.

"Don't I get a receipt?" he sneered.

"Take your shit and go," I said without looking back.

I immediately reported his harassment to T.Sgt. Chipo. He dealt with it. By dealt with it, I mean he laughed and put another tick on the white board.

"Clark, will you just be honest with me," he said. "You love this attention, right? I mean, it must be nice to be *the Beyoncé of deployment*. Every guy out here keeps throwing himself at you. Are you going to tell me you don't love every minute of it? You should consider it a compliment."

"I consider it," using his words, "unwanted sexual attention," I told him straight-out. "He was saying perverted things to me. You're telling me I should like that?"

"Shit, Clark," he shook his head. "Guys will be guys. And you're a beautiful woman. Can't you take a compliment?"

"A compliment is 'You're doing a great job.' Sexual harassment is licking your lips when you see me. So no, I don't like it, Sergeant Chipo. And I'm surprised I even have to say that to you."

I was so pissed at this ongoing bullshit, I spoke with a female coworker about my stalker. I hadn't planned on saying anything, but tears fell out of me, and once I started, I couldn't stop. She informed her supervisor, a master sergeant, who brought up to my command that "something" had happened, and I was struggling.

My first sergeant sat me down.

"I know there's something going on, Sergeant Clark. Can you tell me what happened? What did he do?"

"Nothing," I answered. "It's fine."

"Look, do you really want some asshole who's brought you to tears leaving here and getting decorated for his efforts? I can't do anything if you don't tell me. But looking at your body language, I can tell something happened to you. I just need to hear it out of your mouth, and then it will be over. Then I will handle everything."

This first sergeant was the rare breed of Air Force leaders who wanted to do the right thing instead of the easy thing. But I couldn't say anything. I had spent years trying to run away from the slut-shaming I'd received since early in my career at

Langley. It got even worse in Florida when my chief brought up my past assaults and used them to say I had a trend of "claiming" assault in front of eleven people. I'd been mocked, laughed at, and harassed for so long that I feared opening my mouth, believing I'd be subjected to the same old shit that followed me wherever I went.

So I kept silent. Thank God, my first sergeant didn't let it go completely. Seeing the trauma in me, he gave S.Sgt. Hernandez a no-contact order. He didn't need to know the details to know that Hernandez had been the source of the problem. S.Sgt. Hernandez couldn't talk to me, couldn't come within a hundred feet of me.

That didn't stop S.Sgt. Hernandez completely. He still watched me through the window at work. I would look up to see him standing at the window. But he was far enough away that he never violated the order. Still, I never felt completely safe. By the fourth month of my deployment, S.Sgt. Hernandez's deployment ended, and I could finally breathe again.

At least for a while.

Chapter 26

Unsub

"If your flirting strategy is indistinguishable from harassment, it's not everyone else that's the problem."

~ John Scalzi

O NE OF MY areas of responsibility was the BSS, our base supply store. This is where airmen purchase books, pencils, pens, trash bags, cleaning supplies, and any office warehouse stuff that you would need. Picture a small store. Then shrink it down to a tenth that size. Our BSS was no bigger than half of a single-wide trailer. Even though it sat on the unsecured side of the base, shoppers still must be authorized to purchase anything. It's like a tiny convenience store for military personnel across branches to buy sundries.

One afternoon, S.Sgt. Campbell asked me to run the store for an hour while he taught an Excel class. Even though he reported to me, he was more like a friend. And once I left deployment, he would run things for me.

"No problem, Campbell," I told him. "I'll be at the store in time for you to teach your class."

As planned, he left once I arrived. The shop wasn't the kind of place that got busy. People would trickle in and out. Most people used it to get out of the heat of the sun. When the temperature reached 110 degrees outside, our shop felt like an oasis at a balmy 100 degrees!

During the hour, I had four service members in the shop: two airmen and two marines. While I checked out the airmen, a lieutenant with the marines entered, someone I'd never seen before. He stopped and talked with his fellow marines for a minute as the airmen completed their purchases. Finally, the two marines left, and the lieutenant and I were alone.

"So, what is this place?" the lieutenant asked.

"Well, it's our version of a base service store," I responded. "It's small, but it meets our needs."

"Can take a tour?" he asked.

Well, there's the door, used both for entering and exiting. And here are our five shelves with very few items on them. How's that? I didn't say that, but inside I rolled my eyes. If you couldn't see something inside this shop from the doorway, it's because you were blind or we didn't have it.

"Do you have a letter from your commander?" I asked. A letter means that you are who you say you are, and that becomes your military authorization to shop at the store. "You need that to shop here. If you do, I can give you a tour if you really need one," I continued.

"Okay, I'll come back," he answered. Instead of turning to leave, he licked his lips. "You have a nice frame," he said. "How many times do you hear that every day?"

"Sir, can you please leave now?" I said, standing as tall as my five-foot, five-inch body would allow.

As I mentioned earlier, the shop was as hot as balls. That day, our thermometer read 104 degrees. To endure that kind of heat, I wore cargo pants and an Under Armour T-shirt instead of my military blouse with my name on it. When it gets hot, you wear whatever helps sweat evaporate away from your body as quickly as possible.

"Sir, I'm not going to ask again. Please, leave now," I said with as much authority in my voice as I could possess.

Instead of exiting, he came around the corner and walked towards me.

"Nah," he answered. "I'm not going to do that."

I backed up as far as I could, which wasn't far in this tiny shack, and he came closer still. I yelled once more for him to leave, while I reached for my phone on the desk.

Before I could touch my phone, he came behind me and pinned my arms behind my back. Then he bent me down over the desk. With one hand, he pressed my face against the desk, and with the other he undid the belt on my pants.

"No!" I heard myself yell as I tried to twist myself free. "Get the fuck off me!"

As the encounter went on, through my sobs, I begged, "Please! Please, don't." But I was in such emotional anguish and spoke in such a low whisper at that point— already defeated by knowing that nothing I was saying would make him stop—that I doubt he heard me. Nor cared. I tried my best to sound menacing and intimidating when I eventually got up the inner strength to scream, "Get the fuck off me!" But I felt so weak and helpless, all I could do was whisper pleas as he overpowered me.

Instead of listening, he slid his hands inside my pants and penetrated me with his fingers while I continued to writhe and twist to get away from him.

Then he got mad, slamming my head hard onto the desk.

"You're no fun," he said as he took his weight off me.

I don't know how much time passed. I stayed bent over the desk for a time, tears pooling on the desk, screams stuck in my throat. Finally, I righted myself, pulled up my pants, and sat down at the desk chair facing the wall.

The next thing I'm sure of is that S.Sgt. Campbell skipped up the steps to the store and opened the door.

"As if it's not already hotter than hell in here, I brought us both coffees," he smiled as he stepped inside. I don't know what he saw when he looked over at me, but he dropped the coffees on the desk and immediately crouched down beside me.

"Hey, what's going on?" he asked, spinning my chair to face him.

"I," I began. "I just need to leave."

"Okay, first tell me what happened?" he asked with real concern.

My voice and body trembling, I told him what happened, while his face took on a clouded, angry look.

"But you can't say anything," I told him. "I'm getting ready to return home. I don't want to leave with an open case. I don't want to leave that way. I just want to get the hell out of here and go home."

Deployment bathrooms are pigsties. Put a shower in an outhouse, and it would look and smell like what we bathed in on deployment. More than once on deployment, I had to return to the shower to scrub all over again because the shower curtain brushed against my butt as I stepped out.

But that night, I went into the shower and collapsed on the floor. I didn't care. I just sat under the stream of water until my skin felt raw and no hot water remained. And even then, I felt filthy, defiled. I couldn't wash off what he'd done to me.

I didn't know what more I could do. S.Sgt. Campbell walked me to breakfast, and other airmen walked me back to my tent. I knew I couldn't return to the store. And now I had a new boss, a female named T.Sgt. Young, who replaced T.Sgt. Chipo. She came in when I had two and a half months left before returning home. Before leaving, T.Sgt. Chipo told T.Sgt. Young that she needed to watch me carefully, because I was a "real troublemaker."

I told S.Sgt. Campbell that I was afraid of the new boss since T.Sgt. Chipo had trashed me to her.

"I'm guessing she'll make up her own mind about you," he said. "Just like I did."

"What do you mean?" I asked.

"When I first got here," he told me, "I mean, literally, on my first day, Chipo and Bear pulled me aside and told me, 'Good luck working with Clark. We've all tried to hit that. She's a real ball-breaker.' And look at the two of us, acting like human beings who respect each other. I've never tried to hit on you, and you've never been a bitch to me," he said with a smile. "All I mean is that I made up my own mind about you, because I could see right through that idiot Chipo. I'm thinking T.Sgt. Young will see through his bullshit, too."

We were a small office, so of course, I saw my airmen six days of the week. Additionally, we ate breakfast, lunch, dinner, and midnight chow together. We became family. We cared for one another in a way that exceeded the images on any Air Force poster.

How in the world have I managed to meet the absolute best and the very worst people in the world during this deployment? I wondered. If it weren't for my airmen and S.Sgt. Campbell, somebody else may have received a Green Dot Award for saving my life.

S.Sgt. Campbell was right about T.Sgt. Young. I had little contact with her at first, and I saw that as a good sign. If she meant to cause me trouble, I assumed she would have pulled me into her office immediately. But she didn't. In fact, she didn't call for a meeting until two weeks into her deployment. Chipo had tried to poison any relationship I might build with her by trashing me to her, but once T.Sgt. Chipo left, T.Sgt. Young made up her own mind, just as S.Sgt. Campbell had said she might.

"Hi, Clark," Young greeted me one morning. "We processed new shipments this morning, and the shelves are overflowing. Can you load up the truck and drop these boxes off at the store?"

I didn't feel it coming. It's as if I'd blocked all thoughts of the store from my mind, just like I'd done with SJ since I deployed. But when she asked me to return to the store, a flood of pain rushed through my body, and I broke down.

"I am not," I said, hyperventilating, "Going. To. The. Store."

Then my tears turned to rage, like a demon had taken over my body, using me as a puppet.

"What the hell is going on, Kia?" she asked. "Let's do this," she said, putting her hand out to me. "Let's step into the break room."

Once we sat in a quiet corner, she leaned forward and talked to me in a quiet voice. "What the hell was that about?" she asked kindly. "I've never seen you act that way before. What's going on?"

I had gotten lucky again. Young seemed to have a compassionate heart and insight. I told her what happened with the marine.

"Oh my god," she said. "I want you to know that I would never have asked you to go to the store had I known. Why didn't you tell me?" Her face and body language took on the tension of a mama bear that saw a threat to her cubs, and I could tell she wanted to beat the hell out of the marine.

"T.Sgt. Young," I said after thinking about it. "You are a mandatory reporter. Anything I tell you, you have to report. So I will say again. Something happened. In the store. Involving a marine. And that's all I can say except I'm not sure I can return to the store."

Had the lieutenant who attacked me in the store left me alone after that day, I would have done nothing. Because I didn't even know his name. From watching *Criminal Minds*, I thought of him as the "unsub": unknown subject. But shortly after I met with T.Sgt. Young, I saw him again in the DFAC line for chow. When he first entered, he stood at the end of the line. Eventually, he found someone he knew directly in front of me, so he stood facing me close enough to touch. That's when I saw his name on his uniform.

Lewis. Lieutenant Lewis.

Later, I went to the recreation center full of pool tables, air hockey, movies, and other things for military personnel from all branches to do on their downtime. More than once when I sat there with some friends, Lt. Lewis would come in and make himself known to me. He was a like a dog that would stare at me until he caught my eye.

One night I sat there with S.Sgt. Campbell and another marine lieutenant, Lt. Reynolds, eating Subway at our table.

"Lt. Lewis a buddy of yours?" Lt. Reynolds asked. "I ask because he's staring over here at our table." I sucked in a deep, jagged breath as S.Sgt. Campbell held his hands out in front of himself, patting the air in a calming motion. S.Sgt. Campbell knew from my reaction who must be staring my way.

"Let's just say he's made himself known," S.Sgt. Campbell responded.

"I'm not surprised," Lt. Reynolds replied. "He's scum."

I couldn't have been more relieved to hear that. I ended up telling him my story, the unedited version. Even though he was a lieutenant, making him a mandatory reporter, I believed he would respect my wishes to do nothing.

When I finished, he leaned back in his seat. "I'll take care of it," Lt. Reynolds said.

"No, I don't want you to do anything," I said quickly. "Do not take care of anything. I'm leaving soon."

I would have disappeared from deployment without incident, but Lt. Lewis couldn't help himself. A week before I left for home, he sent me an email. I have no idea how he got my email or knew my name. The day he assaulted me, I had removed my blouse with my name on the chest. From that day on until I returned home, I wore that blouse every day. I didn't care how hot I got or miserable I felt, I didn't want anyone looking at my body again.

The reason I got assaulted is because I took my blouse off and exposed my T-shirt, I thought. *I gave him a reason. It was my fault.*

"Just wondering if you're still here," he wrote.

"I am rotating out, sir. If you need assistance, please contact S.Sgt. Campbell," I replied.

"No, I'm good," he replied immediately. "Just wanted to see how you're doing. I want to ask you something."

"Sir, direct any questions to S.Sgt. Campbell," I replied.

"No," he typed back. "Something I wanted to ask you offline or in person. Do you have WhatsApp or Snap?"

He was asking me if I had an app that couldn't be tracked.

"No, sir, I do not," I responded.

"Okay, then can we meet in person?" he persisted. "And please, don't be so formal. Stop calling me sir."

"Sir, I have no personal business with you, so I don't know what you're implying or why you would want to meet," I typed back. "If you need anything, please contact…"

"No, I don't have any business to discuss. I just want to meet in person, okay?" he begged in his message.

I stopped communicating with him at that point. Fear covered me like a film of sweat. There were two people in the store that day: him and me. *Did he want to meet me so he could kill me, to make sure I could never tell anyone what he had done?* I came into the Air Force thinking the best of everyone. But I had become jaded. And worse than jaded, terrified.

During deployment, I made friends with the commander's executive. She served as the gatekeeper for the commander. To be safe, I forward her the email chain.

She reached out to me minutes later.

"Kia, this is NOT okay," she said. "Who is this man?"

"I'm not overreacting, right?" I asked.

"He should not be talking to you like this," she replied. "This is wrong."

"Thank you for looking it over," I said sincerely. "But please keep this between us. I don't want to deal with this before I leave for home."

Well, she did what she felt she had to do. She sent it to my chief, commander, and first sergeant. They started their own informal investigation.

"What did he do to you, Sergeant Clark?" my chief asked. "Please tell us what happened, so we can deal with it."

If all my Air Force leaders were like the three of these, I would never leave the Air Force. They'd have to retire me at sixty-two when I reached mandatory retirement age. These leaders made me feel valued, even loved. I had finally made a name for myself in a good way. Here at this undisclosed location, it didn't matter what others said about me. All that mattered is how I performed, delivered, and built strong working relationships. I was their go-to person. They had awarded me NCO of the month, something T.Sgt. Chipo always wanted but never received.

So, when they told me they wanted *to protect me and take care of me*, I believed them.

A few days after the assault, I had trouble concentrating at work. People around me noticed I seemed off my game, and I remained so scared that I couldn't sleep at night. I became paranoid and sick. I kept replaying the tapes in my mind—the ones where I'd had two stalkers, a groper, and someone who sexually assaulted me in a matter of months.

As I continued to struggle, my chief made it his business to help. At some point, I told the mental health specialist and the victim advocate every detail. Then something happened. I filed a restricted report about the assault. A restricted report doesn't start an investigation, but it lets leadership know that an assault happened on a base. My name appeared on the report, but the details were marked undisclosed.

Had I made an unrestricted report, all the names would have been released, and an investigation would have started. I believed in my leadership. But my previous experiences gave me serious trust issues. I wasn't ready to go on record.

And I didn't have to. They had enough information without me naming names. They took care of it, against my will, and did a third-party report. Before I left deployment, many people in my command and office watched over me to ensure nothing happened until I left.

I let nothing interfere with the joyful, tearful reunion I had with my daughter and mom! Being back with them made my hell from the past several months disappear. And I kept those demons at bay for nearly two weeks after I got back to Florida, when I received a phone call from an investigative officer.

"Your name has been mentioned related to a case against Lt. Lewis," the voice on the other end of the phone said. "We'd like a little more information, because he's going to be prosecuted."

"I'm sorry," I said, confused. "What did you say this is about?"

Shit.

Chapter 27

Welcome Back to Hell

"Hell is—other people!"

~ Jean-Paul Sartre

"WHAT CAN YOU tell me about Lieutenant Lewis?" the investigator said again. "I understand you had a run-in with him."

"No, sir," I replied. "I don't know anything about any Lieutenant Lewis."

"That's not my understanding," the investigator pressed. "I read in this report that he touched you."

"I repeat, sir," I repeated. "I don't know this man."

Which was true to a point. When he assaulted me, I didn't know who he was. When he stalked me, I read his name off his shirt, *the idiot*. When he emailed me, I saw his name again. When S.Sgt. Campbell and the other marine lieutenant talked about him, I knew we were talking about the same man. But no, I did not know this man, and I did not want to know him.

The investigator's questions started a series of flashbacks, each one more painful than the next.

"All I need from you…" the investigator started.

"I want no part of it, sir," I interrupted, starting to shake. "As I said, nothing really happened. We had a run-in at my base supply store. He looked at me in a weird way and licked his lips. That's all I remember."

But it didn't matter. I didn't know it then, but they were going to pull me in if I wanted to be involved or not.

Another surprise greeted me when I returned to base: a new squadron commander. Soon after I got back to the office, my chief, the same one I'd had run-ins with before deploying, pulled me into his office.

"Your problem with Airman Karen has escalated while you were away," my chief said. "So, what I want you to do is apologize to her. And then I need you to apologize to some other people, too." He fired off a list of names of the individuals who were close to the situation, telling me that they also deserved apologies.

"I've already apologized to her and tried to clarify what I said," I told him. "Her male friend stopped talking to her, and she blamed me."

"I don't care," he shook his head. "Apologize again. Once you do that," he said, "I think we can move past this. Then you might be able to regain favor from leadership."

Whatever.

I apologized to Karen—*again.*

"I'm glad you apologized," she said, "because I was so afraid of you coming back. I thought about cutting your brake line and slicing your tires. But," she continued, "this situation became so public, I think you should apologize to me in front of everyone at the next commander's call."

Not only did I ignore her comments on fantasizing about sabotaging my car, but I continued to apologize until I'd spoken to every name on my list. For my efforts, nothing changed. She still hated me. More problematic, my chief still hated me. I started thinking that he had me apologize just to humiliate me again and try to take the luster off my successful deployment.

But I didn't let that deter me. I determined to keep moving on with a good attitude. I had a fantastic deployment from a career perspective. I'd worked hard and made a name for myself.

My new superintendent put me up for the Logistics Readiness (Log-R) Airman of the Year Award based on my performance results. That prestigious award is earned by one person on the base, so being nominated is an honor itself. Even though certain people hated me, I still managed to be the top candidate on the base. However, my nomination had to be approved by my new commander, someone I'd never even met.

When it reached his desk, the commander stopped it.

Simultaneous to the Log-R of the Year Award, my superintendent and flight commander put me up for Outstanding Airman of the Year (OAY). This is an even more prestigious award. The 12 OAY winners earn the Outstanding Airman ribbon with the bronze service star device and get to wear the Outstanding Airman badge for one year. They also serve on the service's Air Force Enlisted Council for a year. I was thrilled with the possibilities that could bring me. However, once the commander shot down my Log-R of the Year nomination, my flight leadership pulled my name off the list.

"You were my top pick, Clark," my superintendent told me. "I'm new here, so I made my recommendation based on your performance, not what others say about you. But let me be frank. I've learned that others here know something about your past they won't let go of. I don't know who you killed, but I'm stuck in the middle. I caught hell for even submitting your name for that award. I was told that nominating you would be a wasted nomination, so I had to pick someone else. I'm sorry."

"Why does the commander have such a bad opinion of me?" I asked.

"My understanding is that Karen basically lived in the commander's office when you were on deployment," he explained. "And then she'd talk to your chief. She trashed the hell out of you. As the days got closer to your return, she told everyone that she was afraid of you."

Karen—just like S.Sgt. Morin, Kramer, and even my soon-to-be-ex-husband—had the ear of leadership. If I said or did anything that questioned their version of the facts, I was met with a brick wall.

"I need to get out of this section," I told my superintendent. "I'm sure the commander filled you in about the situation with Karen, right? I spoke to Karen about perception and being professional at work, as had three other NCOs before me. Now I'm good for shit as far as the commander is concerned? What the hell?"

Since I needed to get out of my current section and had done so well in my legal studies in high school, I applied to the Offices of Special Investigation, which is like the FBI of the Air Force. I figured I had enough experience as a victim to give me the ability to see both sides of any situation. But before I could retrain, my commander needed to approve it. I scheduled a meeting with him, hoping that he hated me enough to be eager to let me go.

My superintendent called me in a few days before the meeting scheduled with the commander.

"I'm sorry, Clark," he said. "The commander refuses to meet with you to discuss retraining. When he saw your name on his calendar, he gave me hell."

"Why?" I asked furiously. "Isn't that part of his job? How can I get out of here if he won't even talk with me?"

"You're my top NCO, Clark, and you know that I stand behind you," he said with empathy. "He told me that you have a checkered past."

"He's new here!" I responded. "I have no past with him. He just got here while I was on deployment getting decorated. Can I at least schedule a meeting with all my leadership to find out why I'm blacklisted? My chief told me that if I apologized to a bunch of people I'd get a reset on any past problems. That was utter bullshit. All it did was undermine my credibility and force me to eat shit for doing my damn job."

"I hear you, Clark," he nodded. "Let me see if I can schedule that meeting."

In the meantime, my soon-to-be-ex-husband SJ returned from deployment. We would be living under the same roof for a month while we both found new places to live until our divorce became final.

Throughout my deployment, I reconnected with many of the people that SJ insisted I cut out of my life. One of them was James, Harold's roommate back in Virginia, the guy I jumped on the motorcycle with to go with to the grocery store. When I first met James, he was a civilian. He'd since joined the Navy. We reconnected, and I enjoyed his friendship. For the first time in years, I had a male friend in my life, and I valued him above all others.

Once SJ returned from his deployment, I introduced him to James, praying that James would remain in my life even with all the drama circling around me. Since James would be spending time with Reign, I figured SJ would want to know anyone who would be in contact with our daughter.

Shortly after SJ's return, he and I sat in the same room watching TV. He sat on the recliner, and I sat on the couch with Reign. SJ and I kept separate quarters except for two common areas: the kitchen and the TV room.

"I'm really happy for you," he told me. "James sounds like a great guy, and I wish you both the best."

"Thanks," I replied, relieved to be beyond his tears and rage. "We're just friends, but you're right. He's a really great guy."

"I got you a little something to celebrate," SJ said, producing a bottle of my favorite wine.

"Just because we're getting a divorce doesn't mean we can't be civilized. How about a toast to new beginnings?" he suggested.

"That sounds nice," I told him, letting down my guard.

SJ poured me a glass of wine while he opened a bottle of beer for himself. I stretched out a bit on the couch while SJ settled into his recliner. Whenever my glass got below the half-full mark, SJ refilled it while he slowly nursed his beer. I don't remember getting buzzed or tipsy. Instead, I went from sober to drunk. I had been deployed in a country where alcohol was prohibited, so I got intoxicated quickly and without warning. My tolerance, while never high, dropped drastically after going several months without any. My eyes settled on the empty wine bottle on the coffee table.

Shit, I thought as I wobbled standing up. I headed into the kitchen to get some water. Once I got there, I leaned on the counter that had inexplicably moved since the last time I leaned on it.

SJ rushed over to steady me, when he noticed my stumble.

"Okay, okay," I said. "I need to go to bed."

As I turned towards my room, SJ put his hands around me from behind, at first attempting to steady me. Then his hands settled on my face, as he started to gauge how drunk I was. Next, he groped me, pulling my shirt to one side to expose my breasts.

"No, no," I slapped his hands away. "We shouldn't do this. I'm not yours anymore."

"It's okay, it's okay," he assured me. "We're still married."

As I stumbled to gain my footing, he took my hand and placed it on his penis.

"Please, don't," I said to him.

Then I blacked out.

When I woke up in the night, I was wearing different clothes. I felt dirty and ashamed. I didn't remember details, but I knew that something had happened.

I confronted SJ in the morning.

"That was not cool what happened last night," I told him. "You know that I want to be with someone else. You got me drunk and took advantage of me."

"It's okay," he waved his hand. "It's water under the bridge. We've both been deployed a long time, and neither of us have had sex. We got a little carried away. It's fine."

"It's not fine," I sobbed. "You know I want to be with James. I'm not yours, but you took what you wanted, just like you always do."

"Well, don't tell James," he suggested. "You know, what he doesn't know won't hurt him. And I've already told you, *you can't rape your wife. You can only take advantage of her.*"

I called James and told him everything. Not only was James understanding, but he said something else I needed yet didn't want to hear.

"You were assaulted, Kia," James told me. "You need to say something."

"He's the father of my child," I cried. "I hate him so much, but I can't send him to jail. And I can't do anything while I'm still living here, because I'm afraid he'll kill me."

"Then don't do anything right now," James advised. "You need to be safe. Maybe it's time for you to move out into your own place. If you don't have the money, I'll send you money to break your lease and get you into a new place. But I'm worried he's going to keep assaulting you if you stay there."

The last night SJ and I stayed under the same roof, he wanted to talk.

"I know that I'll never find love again," he said in his pity me voice. "I mean, who would want a single, thirty-four-year-old dad?"

That's when I realized that SJ was working on his next character, SJ the Forsaken. He was practicing his divorced, single-dad lines on me. I could hear him working out the pitch he'd use on his next victim.

"I'm just a poor, single father wanting nothing more than to see my precious daughter. Unfortunately, the wicked baby mama won't let me see her. I just lost my own father, and then my wife leaves me after I gave her everything. My heart is so broken. Why have they forsaken me?"

I imagined that he would return to bodybuilding to get buff as an extra attraction on his lure. And that's just what he did.

I moved out in December. James came down to help Reign and me move into a new place. Then James confronted SJ about what had happened.

"Yeah," SJ said, lowering his head. "I fucked up. I'm sorry."

He couldn't even look at James. SJ had always been non-confrontational with other men, unless he held extraordinarily strong cards or communicated via text. He preferred to bully submissive woman. Instead of playing the alpha male, SJ acquiesced and kept his head lowered.

After that conversation, SJ kept his distance whenever he knew that James was around.

True to form, SJ kept refining his new persona. It wasn't long before he realized what a valuable prop our daughter could be in his little play. He pushed for parental rights, believing that façade made him a hero and allowed him to "save face." Besides, what could be a better chick-magnet than a Facebook profile photo on Father's Day or a sappy Christmas post about how he would give anything to be in her life every day if not for "wicked, old baby-mama"?

As hard as it had been for me to get away from SJ, I still had fiery hoops to jump through to protect Reign. And I knew that I'd have to work just as hard to get away from my commander in Florida if I had any chance of becoming whole.

Chapter 28

Jingle Bells

"God moves in a mysterious way."

~ William Cowper

I N LATE NOVEMBER, my superintendent confirmed a meeting with me and my commander, chief, first sergeant, superintendent, master sergeant, section chief, and flight commander. In other words, my entire Air Force chain of command would be in the same room with me to tell me what I needed to improve to get into retraining.

Before the meeting, my supervisor called me for a pre-meeting meeting.

"So, do you have any idea how this is going to work?" I asked him.

"No idea," he shook his head. "I'm happy to stand up for you, but is there anything specific you want me to say on your behalf?"

"Thank you," I responded. "Would you just tell him what you see in me? If I'm a good worker, tell him. If I'm a good leader, tell him that. Just fight for me, even if you strike out."

After entering the commander's office, we all stood behind our chairs at attention until he spoke.

"Have a seat," he spoke as a formality and a way of officially beginning the meeting.

We sat. No one said a word. Eventually, I figured out that since I'd requested the meeting, it fell to me to start.

"I know how valuable your time is, sir," I began. "As you know, my name was pulled from both the Log-R Award as well as the Outstanding Airman of the Year. Later, my request for retraining was pulled as well."

I made sure not to speak more directly. I chose "my name was pulled" instead of "you pulled my name" to avoid any finger-pointing.

"I graduated from ALS (Airman Leadership School) as a distinguished graduate. Before leaving for deployment, I saved a life and was recognized at a coining ceremony by the wing commander. I was selected to conduct an interview and shoot a video to reinforce the value of Green Dot training. While deployed, I was recognized as the NCO of the month," I continued. "I mention these merits and service distinctions to demonstrate that I've been a solid performer, both in my job and as a leader. The reason I requested this meeting is to learn from you what I've done that would make me ineligible for retraining, sir."

"Nice speech, airman," the commander replied.

Uh-oh. He just called me a bitch, I knew from experience.

"But as long as I'm sitting at this desk," he said, making a show of pointing at the top of his desk, "it doesn't matter what you do. You will never get retraining. I will not sign off on anything that had your name attached to it."

"I don't understand, sir," I struggled to say without releasing tears.

"You will, Clark," he continued. "Do you know why you're stuck here? I had an airman in my office every day in tears telling me how scared she was that you were returning to base. You know what that tells me? It tells me you're a toxic leader, an evil person, and I don't like people like that. In my eyes, you will never succeed, at least not by my hand. So, if you have the power to build your own career by yourself, have at it. But I will never lift a finger to help you."

Evil person? Toxic leader? I wanted to be angry, but instead I felt more hurt than almost any other time in my life.

I said nothing, hoping that someone—anyone—else in the room would speak up on my behalf.

Crickets.

Eventually, the commander broke the silence.

"OSI (Office of Special Investigations) is full of professional," he punched the word for emphasis, "men and women with qualities you will never possess. OSI is a career field where they investigate people like you, and I won't have my name attached to a recommendation. Do you really want them asking me what I think of you?" he asked, leaning back in his chair, unmoved by my words or the accolades I'd received from other leaders. "Airman, you're not ready to move on or move up."

"Okay, how much more time do I need?" I asked, swallowing a couple of times to force the words out of my parched throat. "A couple months? Can we revisit this conversation in a month or two—or a year? Can you give me an idea of when I can come back to talk to you about this?"

"There is no specific time. But I know that time will never come while I'm sitting behind this desk," he said in a way that reminded me of Lucy pulling the football away just before Charlie Brown began to kick it, ensuring that he'd miss every time. "Do you know what a *running record* is? I was briefed about you when I assumed command, and there is nothing that you can say or do that will ever make me change my mind. I do not care how many lives you 'save.'"

I knew that his mention of my running record was his way of saying that if I remained at this base, he would inform all new leadership about me. I would never escape his judgment or reach as long as I stayed at that base.

"Do you mean *never*?" I asked desperately. The tears were coming, and I felt my teeth clamping down on the insides of my cheeks to keep my body from breaking into pieces with sobbing.

"Sir," my supervisor broke in. "I understand your position, but if we're trying to mold people and help them get better, is there a reasonable time frame where we can check in, maybe three months from now, to see how she's doing and if she's making progress?"

"I'm not giving a time limit, and I doubt any amount of time will be enough," the commander said quickly.

More silence across the room. I closed my eyes and prayed that someone would tell me I wouldn't be deemed as unclean for the rest of my military career.

"Um, Sergeant Clark," my flight commander said to me, "look, you're doing all the right things. Your job performance is great. We've heard no complaints. Just keep doing what you're doing."

"Sir, I don't understand," I said, totally confused. "You're telling me that I'm doing great and keep doing what I'm doing. But the commander just said that it's not enough, and it will never be enough as long as he's at that desk."

"Look," the commander backtracked slightly, "I'm not saying you're irredeemable, Clark. But as far as any good thing that others in this room may see in you, I just don't see it."

How could he see everyone in every flight? I wondered to myself. *I thought that's why the military had a chain of command. Aren't the other leaders around the table his eyes and ears? If they say I'm doing a great job, shouldn't he listen to them?"*

"If you have nothing new to add, airman, you're dismissed," the commander said.

We all stood at attention, per protocol, remaining in that position until he officially closed the meeting.

"Carry on," he said, signifying we were done.

As we walked out, my supervisor caught up with me.

"Clark," he said, "that didn't go the way I hoped. But like I said in his office, you're doing a fantastic job. I wish I had more NCOs like you."

"Why?" I argued. "So they could be hung out to dry by the commander, too? If I keep doing what I'm doing, I'm gonna keep getting what I've gotten—a pile of shit and a stalled career."

"I don't understand it, either, Clark," my supervisor responded. "I'm not sure what was relayed to him, but maybe he thought what you were accused of was the same sort of thing OSI investigates. We don't know what he was told about what happened, because to him it was much worse than a DUI would have been."

"If I'm so irredeemable," I told my supervisor later, "why haven't I gotten paperwork or an Article 15? Why did I make rank? Why have I continued to receive awards and decorations? And now I'm wondering why the fuck I'm even still in the Air Force," I said, as my hurt turned to anger.

But this wasn't my only disappointment and heartache. I kept getting calls from the investigator looking into Lt. Lewis, and I had just moved into a new place after SJ sexually assaulted me.

The only constant in my life became my friendship with James. It didn't matter who else thought I was a fuck up. Next to my mom, James became one of the few constants in my life.

By this point in my life, I had a lot of experience with putting on this mask. My depression took on the shape of a smile. No matter what was going on in my world, I put on a happy face.

For all the good it did. No one talked to me. They put me in a section all by myself with no other coworkers and no airmen to supervise. My superintendent stated that the commander had ordered that I supervise no one to intentionally ostracize me. So I continued to work in silence while trying to remain hopeful.

December rolled around, and I looked forward to the base Christmas party, but not for the party itself. I got excited because James would be joining as my plus one!

As the party began, I quickly worked my way over the snack table. Old habits die hard. My excitement at a table full of goodies got cut short when the commander walked over and stood next to me.

"Airman," he spit out like a dirty word. "Yup," he said next as he rolled his eyes and walked away.

How nice, I thought. *My own personal appetite suppressant.*

Not long after that exchange, we played a game called human obstacle. The object of the game was to run through an obstacle course with a spoon in our hand balancing a ping-pong ball. James and I joined in on the fun. The commander played with a little too much zeal, knocking into players to force them to drop their balls to the floor.

"Oh my God," James said to me. "He's really drunk. Did you see him knock that guy to the ground?"

I'd been to enough military parties to see lots of stinking drunk people. I hate to say it, but alcohol is the most misused drug in the military. But I had never seen a commander get wasted in public. It's just not done. Usually, commanders have a drink or two all night, if any at all. Basically, commanders make an appearance and spread a little cheer to the troops. Not this guy. He seemed pig-drunk, and his behavior would have had any bouncer in the civilian world toss him right out the door. If this were a frat party, the commander would have been doing keg-stands on the stage.

Next, they played a game that combined musical chairs with *Let's Make a Deal*. James and I sat this one out. Participants sit in chairs facing the audience. The person on the mic yells out an item, like "Chapstick!" Then the participants leave their chairs, run out into the audience, then try to find the item the quickest. The last person to secure the item is out.

While people played, the commander blocked the way of participants when they tried to run into the audience. No matter which way participants tried to cut, he stepped in front of them.

"What the hell?" James said as he watched the spectacle.

"Welcome to my world, James," I said.

"He's really drunk and acting stupid. Talk about unprofessional," James said quietly.

"Ironic, right?" I laughed, thankful we sat the game out.

The commander wasn't the only drunk person at the party. An airman named Warren got sloppy-drunk fast. I remembered her from the previous year's party when she tried to feel me up. That party was on a boat, and Warren swayed so much I thought she might fall overboard.

Good times.

"Um," I said to a sergeant, "do you think it's time to cut off Warren before she gets hurt?"

"Fuck no," he laughed. "That's just Warren. She does this every year."

In my way of thinking, *if Warren does this every year, why hasn't anyone pulled her aside for counseling? What if she gets drunk and hurts herself or someone else? But, hey, what do I know? I can't even counsel my own employee without getting a target drawn on my head.*

James and I stood at a cocktail table, me having a drink, and James nursing the same bourbon most of the night. Since this was James's first time around my squadron and he didn't know anyone, he played the observer. He didn't know what kind of people were at the party, and we wanted to keep his head clear and mind sober in case we needed to get out quickly. It wasn't long until he started to know more about this crowd than they would have liked them to know.

"Look over there," he said, nodding his head ten feet away. I looked and saw Warren knock into a table, stumbling to the point where she nearly fell to the ground. As an officer reached out to steady her, she ripped her arm back away from him hard. The music kept me from hearing the exchange, but it didn't look friendly. She looked like she wanted him dead.

Deck the halls with boughs of holly...

Warren was a car accident I couldn't stop watching. Unfortunately, her drunken eyes found mine, and she took a couple of stumbling strides my way and joined us at our table.

"Did you see that fucker?" she slurred. "He had his fucking hands all over me. Fucker."

Fa-la-la-la-la, la-la-la-la...

From what I saw, it looked like she fell, and he tried to help her and see if she was okay. *Not my circus, not my monkeys.*

While Warren talked to me, she kept weaving in and out of my face. I assumed since her eyes wouldn't focus, she moved her body to try to see me clearly, just like farsighted people move whatever they're reading closer and farther away from their eyes to focus.

"Are y'all friends of hers?" I said to a couple of airmen who joined her at our table. "Because one of you needs to call her an Uber, like, now. Get her home. She's ready to black out."

I don't know if her friends answered me or not, because Warren went from talking in my face to touching me.

"Girl," she slurred, "you look like a chocolate Jessica from *Who Framed Roger Rabbit* with that red dress you got going on."

"Okay, that's enough," I said, removing her hands from me. "We have to go now."

James and I left our table and roamed to a new spot in the room, trying to find less inebriated company. Eventually, we started dancing, forgetting about all the drama for a time, and just enjoyed ourselves. Then James sat one out and stood next to nearby table while I danced by myself. I'd forgotten how much I loved dancing, since SJ would never let me do it. Like many things, I'd told myself at the time that I didn't enjoy it anyway. But now that I had a friend who liked dancing and didn't keep me from it, I felt like a stringless puppet for the first time in years.

Pretty soon, I felt someone grinding on me from behind. I turned to find Warren. Not only hadn't she left the party, but she had found me again.

"Nah, girl," I said, pulling her hands off my body. "We ain't doing that. Keep your hands to yourself."

I got pissed, but I had been around enough drunks to know the only way to deal with them is if one of us left the party. I left the dance floor and followed James to a new table.

"Did you see that?" I asked him.

While the two of us stood talking, I felt two hands reach around me from behind and go up my dress. Then I felt fingers on my vagina.

I spun around.

"What the fuck?" I screamed, turning to see Warren giggling.

Instead of letting go, she groped every part of me she could reach, until James gave her a gentle push to put distance between us and her. Then someone else who'd seen what she did gave her another push to send her further away.

I don't remember who said it to me, but someone who'd been near enough to see it all happen said to me, "That's Warren. That's just what she does!"

You shittin' me? That's like excusing Jeffery Dahmer by saying, "That's Dahmer. It's just his thing. He kills and eats people."

"I need to say something," I told James. "I'm going to the commander."

"Let's step outside so I can hear you," he said, walking us outdoors.

As soon as we got outside, I saw my commander. Before I had time to second-guess myself, I walked up to him.

When he recognized me, his face seized into a look of disgust that said, "What the fuck do you want now?" I froze up before I could say or do anything. Before I could speak, he rolled his eyes and turned away from me.

James and I stepped back inside the doorway, where I spotted the commander's wife. I wanted to speak to her, but then I thought to myself, *if I can't even face my own commander, what makes me think I can talk to his wife, a civilian, with no power to help?*

Instead of speaking, I stood there with my mouth open like a dead catfish.

Chapter 29

Utter Hypocrisy

"It takes courage to grow up and become who you really are."

~ E.E. Cummings

O N MONDAY, WE returned to work, but I could not get what happened out of my mind. In a matter of months, I'd been sexually harassed by more than one NCO. According to T.Sgt. Chipo's tally on deployment, I "earned" twelve tick marks in the three months Chipo was there, and who knows how many after he left. I had been stalked by two other NCOs, penetrated by a Marine officer, sexually assaulted by my soon-to-be-ex, and now sexually assaulted by a woman at an office party. I knew little about post-traumatic stress disorder, but my typical coping mechanism of putting on a big smile was failing me. My thoughts were distorted like a fog bank setting into my head, and they wouldn't leave.

During the first thirty minutes at work, I started to die inside. I didn't label it as PTSD or depression, but I knew that I couldn't live like this for long. Finally, I reached out to an SNCO who had moved from operations into Green Dot training. So not only was she a friend and confidante, but she also ran the program to help reduce episodes of suicide and sexual assault.

We connected by phone a short time later. And I spilled my story. I mean, I told every detail I could remember. When I finished, I felt some relief having unburdened myself.

"Kia," she said slowly, "I know what I'm about to say is going to sound super hypocritical, especially coming from me, but Warren is like that every single year at the parties. I think she probably has some stuff going on at home, and she really lets loose at holiday parties. Do you know her? She's like this quiet, even mousy, person who would never hurt a fly. I'm sure you've gone out with the girls before, had one too many, and things went sideways. Girls will be girls, right?"

"You're telling me to let it go, do nothing?" I asked, more shocked and confused than ever. The Green Dot program exists to prevent more victims, not protect more predators.

"I'm not saying do nothing," she clarified. "But I think you should go to her, talk to her, woman to woman. Tell her she messed up and you're pissed. Then leave it alone."

"Leave it alone?" I repeated, closing my eyes.

"Because if you report it," she continued, "you're going to mess up her career. She will lose her job. She has a kid and a husband. And she's the sweetest person when she's not drunk, I swear to you! You can't tell me that you've never gotten blackout drunk and done things you don't remember or that you regret later."

"Well, I never got shitfaced at a Christmas party. I never physically or sexually assaulted anyone," I said, annoyed with myself for believing I might have someone looking out for me for a change. "So no, I disagree with your premise."

"Kia," she pleaded. "You're not vindictive, I know that. But this charge is serious, and you will seriously fuck up her life if you report it. And I know how much of a hypocrite that makes me for saying this, but I think you need to talk with her and be done with it."

"You know, we wouldn't even be having this conversation had I been groped by a male. Why is there a double standard? Women can sexually assault both men and women," I told her, shaking my head as I talked into the phone.

Girls will be girls, according to what she just told me, *was more acceptable than boys will be boys?* No, I adamantly disagreed with her premise.

That conversation left me furious and hurt. I thought about Warren. Then I thought about my own stalled career. After sitting with my thoughts for a while, I emailed Warren.

"You in today? I need to talk with you about something," I wrote.

"Yes, I'm here. What's up?" she responded.

"Do you remember anything about the Christmas party?" I asked.

"Yeah. Well, I don't remember any of it except getting really drunk," she answered. "Did I do anything?"

"Never mind," I typed and ended our back-and-forth.

Everything felt so fucked up. The Green Dot instructor wanted me to let it go, but she asked me to confront my abuser. *Do counselors out in the real world ever ask victims of rape to have a little chat with their rapists?*

I tried to sift through more periods of mental and emotional fog, but I didn't feel like I could—or should ever have been asked to—confront Warren. I had enough problems of my own to deal with. I hadn't gone to the party looking for trouble. I just wanted to enjoy time with James. But I got sexually assaulted in the process.

Finally, I summoned the courage to talk with my supervisor. I told him what happened at the party. Then I told him what the Green Dot instructor told me to do. This time, I didn't feel unburdened; instead, I felt betrayed, embarrassed, and unsupported.

"Jeez, Kia," he said, "that's crazy. I'm so sorry to hear that happened. How are you feeling?"

"Sir, I just got sexually assaulted at an office party," I told him as calmly as I could. "And no one seems to care. You and every member of leadership tells me to 'use the resources at your disposal.' But what do I do when the resources tell me that they have more empathy for the perpetrator than the victim? You asked me how I feel. I feel betrayed, angry, hurt, alone. That's how I feel."

"Yeah," he said, letting out a long sigh. "This is a tough one."

And that was the extent of his support.

I called another friend. By now, numbness started giving way to anger. I repeated my story, including the blow-offs I'd had from Green Dot and my supervisor.

"What do you think?" I asked, hoping I'd eventually get an answer I could stomach.

"I hate to say it, honey," she said, "but I think the Green Dot trainer is spot on. You need to talk to Warren. Talk to her woman to woman. Because I'm telling you, she does this every single year. You want to ruin her career over this? If someone did that to me at a party, I'd hit her so hard she'd miss two months of consciousness. You should have put her in her place right when it happened."

"My commander hates me," I answered. "He's looking for an excuse to drop on me like a vulture. If I punched Warren, I'd be the one out of the military, and I'd probably spend time in prison, too."

"Look, I'm telling you what I would have done," she said. "If you had, you'd at least get some justice. Now it's just your word against hers."

I can't win. Had I taken matters into my own hands, I'd be disciplined. I could picture my commander writing up a recommendation for my court-martial or Article 15 documents while saying, "You should have used the resources at your disposal, airman."

In two hours, I'd talked to three leaders in the Air Force. One told me to confront my abuser. One told me nothing. One told me what I *should have* done. But no one gave a flying fuck about me, the victim.

I pride myself in being both calm and rational, so I took my time to process everything before deciding what I should do. And one thing kept coming back to me: "She does this every year."

Every. Damn. Year.

I couldn't let go of these thoughts: *Who's talked to her? Who knows that she has problems at home that drive her to drink so much? Who knows if she has a more serious drinking problem? Who knows when she might go so far that she kills herself or someone else?*

Finally, I reached the *screw it* stage of problem-solving, and I went to see my first sergeant.

"Hypothetically," I started, and I didn't stop until I'd finished sharing my *hypothetical* problem with her.

"Did something happen to you, Clark?" she asked.

"No, ma'am," I answered. "It's a hypothetical."

"Well, she started, "then hypothetically, you know the right answer. You know what's right. You know that if something happened, you need to report it. What's standing in your way?"

She listened as I told her about my terrible interactions with the commander, and why I believed that no matter what I did, I would be the one the Air Force put on a cross.

When I finished, she took me to the office of the chaplain.

Before we went inside, she asked, "Have you ever spoken with a chaplain in the Air Force before about something like this?"

"Yes," I told her, "I went a few times with SJ, but I couldn't speak freely."

"This time, whatever you talk about stays between the two of you," she said, showing real compassion. "Base chaplains can't report sexual assault. Hell, you can tell him that you murdered someone, and he can't say a word. You are in good hands here, Clark."

After talking with the chaplain, it felt like I got more clarity on what I should do. At the very least, I knew that if all else failed, I had the support of the chaplain for help. I needed to report the incident.

I told my first sergeant I would be going to the SAPR office. After she approved my absence at work, I picked up James, who was still in town, and he came with me as moral support as I filed my report. While I was willing to file, I made it a restricted report. Again, restricted means no names.

"I just want to make sure," she said. "You want to make this restricted? It happened at a Christmas party, in full view of everyone. Your commander was there, and you said he was drunk. This is a huge problem. I will do what you wish, but if you file a restricted report, there's no way to make this vanish without an investigation."

I remembered a time when I had been this vivacious, happy person with a quick smile and kind words for everyone around me. Then I'd been emotionally, physically, and sexually beaten. What remained of me was more shadow than substance. But I needed to reclaim myself, to learn to stand up for myself again, even if that meant enduring more torment. And I needed to do the right thing. Warren could be exactly like me: afraid of her own shadow, struggling with mental health issues, and needing an intervention to save her from self-destruction.

"Do it," I said after reminding myself that this could end up a win-win for both me and Warren. Maybe Warren would get the help she needed, like treatment for her drinking problem. Over time, she could recover her lost reputation as *that airman who got loaded and stupid each year at the holiday party.* Even though her antics became a standing joke, not a single airman offered her help. *If those who counseled me against taking any action really cared so much for Warren, they would have gotten her help,* I thought.

Later that day, investigators came to my house, collected my clothes for DNA evidence, and started conducting interviews of everyone who might have witnessed the assault.

My case met monthly at a review board meeting that included Security Forces, OSI, mental health, the vice wing commander, command chief, SAPR representative, the unit commander, and first sergeant. At these monthly boards, they give and receive updates about each open case and where things are in the investigation and legal process.

At one review board, my squadron commander said something revealing.

"Well," he said with a shrug of indifference. "Clark has a history."

"So, what you're saying, just so I understand you correctly," my biggest advocate challenged him back, "is that she deserved to be assaulted, because she has a history?"

Let me skip to the end. Warren received some sort of discipline. I wasn't privy to the details, because her discipline had nothing to do with her sexually assaulting me. Warren got disciplined for what she did to the officer who tried to help her up. Apparently, getting into a verbal dispute with an officer was more grievous than sexually assaulting a mere airman. Either way, I took solace. Her behavior got put on record. She would think twice before doing it again. And I'm sure she was offered help for her drinking.

And finally, by taking a stand, I reminded myself that I deserved better than this kind of treatment.

Before the investigation wrapped up, though, I requested to be removed from my unit due to the ongoing mistreatment I received from my commander. When he told me, "I don't care what may or may not have happened, that's what the investigators are for," I lost any remaining hope that he would ever treat me as a human being. I was placed in the base chaplain's office until they could find me a new temporary place to work.

Because I changed my report to unrestricted, I had many options to choose from, including taking an expedited transfer to another base. This transfer typically takes place within thirty days.

Except, of course, something screwed up my departure.

"You know, Kia," my advocate informed me, "until the Warren investigation wraps up, your commander has the option to reach out to your new base and disclose all the details about your assault. And based on your history with him, I wouldn't be surprised if he didn't share much more, if you know what I mean."

Instead of taking advantage of transferring out within thirty days, I waited as a way to prevent my commander from telling my life story to my new commander.

This delay meant I'd be stuck in Florida until May. In the meantime, I still needed to find a new base and finalize my divorce. Though I'd filed in January, when SJ changed his mind from joint custody to granting me sole custody, the judge delayed hearing our case for another month. Finally, in April, I had my freedom from him.

Except we shared a daughter. Even though I believed SJ to be a shit husband and father, every child needs to believe they have a father who loves them. I did everything I could do to give SJ access to Reign, hoping that he'd want to take an active role in her life. Despite her young age, I could see how our separation was weighing on her heart. I asked SJ if he wanted to transfer to the same base. He agreed to relocate, but we disagreed over where to relocate. Then he returned to vintage SJ, trying to drive my transfer location around his convenience and preferences.

Fortunately, my advocates heard his manipulation and put a stop to it.

"Kia's the victim of sexual assault here, not you," the advocate stepped in. "She gets to go where she wants to go. That's not up to you. If you choose to go elsewhere, go. But she's driving this."

James became stationed in Hawaii, and I wanted to follow him there. Since Hawaii wasn't an option, I chose a base in California to put me someplace warm and as close to James as I could get. I figured if I spent a year at this base, I could put in for a transfer to Hawaii if things got more serious between James and me.

Despite how hard I tried to keep my relationship with SJ solid for Reign's sake, SJ and I found ourselves in conflict again as soon as he started dating. Before I moved, he kept showing up at my house unannounced, and his new girlfriend vandalized my car. More than once he tried to take Reign from my mom when I wasn't home. I felt so frazzled and unsafe, I moved into a hotel until I left for California.

Thanks to the sexual assault program, I had the option to retrain instead of staying in supply. My mental state deteriorated, and I couldn't stay in my career field. I hoped retraining would get me out of this place where I'd become increasingly depressed.

My advocate and I met with my chief and first sergeant, requesting them to get me a meeting with the commander to sign off on retraining.

"Your scores aren't that high," my chief said, referring to my test scores on the ASVAB, the timed aptitude test taken before entry into the military.

"Sir," I corrected him, "my scores are high enough to qualify me for Office of Special Investigations, which is where I'd like to retrain."

"Yeah, I don't know," the chief told me. "Why don't you become a cook? Or security forces? Something where your score won't be held against you."

The chief's recommendations took me backwards in my career path, not a lateral or step up.

My victim advocate spoke for me, and she was fabulous.

"No one is asking you for a freebie, chief," she explained. "What we are asking for is to do what is in the best interest of the victim, Sgt. Clark. She needs a time of healing. She's not asking for a paid vacation."

"Fine," the chief said as he pushed back from his desk. "I'll get you that meeting. Just don't get your hopes up. I'll also speak with the command chief, since I know her very well."

To say the least, I tried to keep my expectations low as my advocate and I entered the office of my commander for our appointment. We made the same arguments to the commander. After sitting quietly while we made our case, he leaned back.

"Do you remember what I told you back in November, Clark?" he asked.

"Sir?" I said, not sure what he meant.

"I told you that I will not sign off on anything, remember? Nothing has changed since then," he said sarcastically. "Let me remind you of what else I told you. If you can get there without my help, knock yourself out. Go for it. But I'm not changing my mind about you. I'm not signing off on anything to help you. Clear?"

I thought to myself, *Do you remember the conversation you had with the vice wing commander and command chief where you agreed to support me in every way possible? Were you lying, sir?*

He leaned back in his chair. "I'll support you in your request for a transfer. Nothing else."

Steam rose off my advocate's head as we exited the commander's office. She was a woman of God, a praying woman, and she looked like she was ready to start throwing fists.

My commander supported my transfer as promised, and he blocked my request to retrain, also as promised. Breaking one of the SAPR program policies, he contacted my future commander in what I believe was an attempt to destroy my character and keep this "running record" he'd held over my head months before. The investigation portion had already closed out, giving him no reason to contact my new commander. Without notifying the other board members beforehand, the commander acted on his own.

When I went to collect my military record as part of my out-processing checklist, my first sergeant refused.

"Clark," he said, "I will be mailing over your entire personnel contents to your new base. I made you a copy, but I will not be giving you the originals. In case you act up over there, they'll have a record of everything that's transpired here."

As I began making the mental transition to California, my anxiety was sky-high knowing that my commander shared not only my entire military history to my new boss, but his own personal feelings about me as well.

This could break me, I thought as I headed to the West Coast.

Chapter 30

Standing Up Against Abuse

"The most precious light is the one that visits you in your darkest hour!"

~ Mehmet Murat ildan

BY THE TIME I arrived in California, I felt as useless as a box of lint. My mind spun out of control, and I no longer had the strength to even fake a smile. Instead of staying in my job and failing, I asked to be temporarily reassigned to the office of the chaplain, where I stayed from June until August of 2019.

I started understanding fully why active-duty service members have such a disproportionately high suicide rate. My mental health plummeted, and my thoughts scared me. Have you ever been in a room with a strobe light flashing on and off? It's disorienting. That's how my brain felt all the time. I'd get a nanosecond of clarity, then I'd be plugged instantly back into utter darkness. I didn't trust the military psychologists. I'd heard too many stories of struggling airmen asking for help, only to have their careers destroyed and their mental states go from bad to worse.

I remembered a line from a Stevie Smith poem, and one recurring phrase kept popping into my head: "not waving but drowning." When you see someone alone and far from the shore, sometimes you need to take a second look. Is he waving? Or is he drowning?

For years, I put on the happy face, and for years, it worked. Whenever someone hurt me, I'd tell myself, "That's on you, not me." Then I'd forgive—and even force myself to forget the transgression. But I found myself beyond smiling, forgiving, forgetting. I'd been sinking deeper and deeper, and no one around me could tell the difference from my fake, happy wave—and me slipping underneath the surface for the last time.

My sister, Brandy, offered to pay for private therapy, something I desperately needed.

From June of 2018 to June of 2019, I'd been sexually assaulted by a marine officer, my soon-to-be-ex-husband, and a fellow airman. The case against Lewis was still an open investigation. My divorce from SJ, finalized in April of 2019, had nearly killed me. And the sexual assault from Warren went nowhere, since she'd committed an offense against someone of a higher rank that same night. I felt invisible and voiceless, except for the few times someone saw me long enough to abuse me. I found myself in a death-roll—spiraling deeper into self-doubt and self-loathing with each breath.

On top of that, I'd begun to despise the Air Force. Yes, I'd worked with some exemplary men and women. But in my current state of mind, I suffered from *negativity bias*, where the only things I could remember were the most recent and painful events and people, one more hurtful than the next. While I was surrounded by sunny, happy people in California, isolation and anguish clutched at my chest.

One light in my life remained: family. My mom, There Daddy, and my sisters stood with me. Reign, who'd just turned three, kept my hands busy and my heart full. And James. James had moved from being a friend to my boyfriend.

Since these people constituted my whole life, they merited more from me than just a woman who drew predators like soda pop attracts wasps. They deserved better. I deserved better.

My Air Force career seemed finished. The longer I stayed, the more used up I was becoming. I decided it was time to leave the service. I began researching and filling out forms to voluntarily separate from the military.

But one ray of hope caused me to rethink that plan. I had met my new commander, Commander Daniels, once. Then I determined to keep my distance, knowing that my former commander had contacted her, which he boasted about proudly. Then one day, Commander Daniels asked me a simple question: "How are you doing?" And I broke into tears.

"I'm so sorry," she said, keeping space between us. "I didn't mean to make you cry. You're going to do well here. We have a very safe and warm community here. We'd love for you to come back to our area once you feel ready to leave the chaplain's office. If you don't think you can, I understand. I won't push. But I want you to know I'm here to support you."

"Thank you," I managed to say, feeling supported, which felt amazing. And in that moment, those words were exactly what I needed to hear.

Commander Daniels helped me transition, and she gave me time to get control of my mental health. She also stood behind me with retraining.

"As long as you put in good work and stay on the right path, you will not get any resistance from me, your chief, or your first sergeant," she told me.

"You should know that I had some bad experiences with my last commander," I told her one day.

"I don't need to hear it," she said with a smile. "He told me himself."

"But I think you should know…" I started again.

"We both know that he called me," she read my mind. "But I give everyone a fair shake. A clean piece of paper, and I let you write your own story. I already forgot what he said, because it doesn't matter. What matters is that you are here, you are safe, and I am more than happy to see you return back to the unit, whenever you are ready. We will be lucky to have you in our unit."

"Again, thank you," was all I could manage to say.

"I don't make up my mind based on the opinions of others," she said easily. "I've forgotten everything he said to me. I will form my own opinions based on what you show me. Take that blank page, and start writing your own story."

Besides Commander Skyler when I'd first started, Commander Daniels was the other outstanding, compassionate, unbiased leader I had. She never judged or pre-judged me, and I never felt anything negative from her.

And I needed her support more than I could have guessed.

Once I felt emotionally settled in California and started to build new, healthier routines, I got an email from the prosecutor handling the Lt. Lewis investigation.

"The good news is," he started, "that we have a few other witnesses. Unfortunately, witnesses mean victims. You weren't the only one. The bad news is that you're by far our strongest witness. We really need you to testify."

That night, I told James I was getting pulled back into the Lewis assault case.

Partially thanks to James's support, I scheduled a phone meeting with the prosecutor so they could fill me in on the case.

"It comes down to this," the prosecutor told me on the call. "If you don't testify, he walks. It's as simple as that. We need you. Otherwise, he's going to do it again."

"Nope," I said immediately. "I will not testify."

"You should know," he told me, cinching my fate, "at this point in the investigation, you can either testify willingly—or under a subpoena."

Story of my life. There were other victims of S.Sgt. Morin, Kramer, and Warren, but I was the strongest one in those three sexual assault cases. Those three cases were investigated and tried by the Air Force. The Lewis investigation and prosecution, though, was handled by the Marines. But one thing remained constant: I had to step up, like it or not. To top it off, I later learned that Bear, the man who groped me the first week I was deployed, went on deployment after an incident with a female coworker back on his base. His victim had been scheduled for deployment, too, but she became so emotionally damaged by his sexual harassment that she couldn't go.

These are the ones I know that have offended more than one person. How many others are out there? And—I couldn't shake this thought—how many more might they assault if I keep my mouth shut?

"Okay," I said in resignation. "Then I guess I'll testify."

Months passed, and the prosecutor kept me on speed dial, keeping me informed on what was happening and when the hearing would be scheduled.

"We have a key witness that recalls everything that you disclosed after the event with Lt. Lewis happened," the prosecutor told me, "but there were some discrepancies with your initial interview back in August 2018."

At this point he asked why I hadn't disclosed everything that happened to me.

"The investigator called as soon as I got back from deployment," I told him. "And frankly, I didn't want to deal with the situation. Then I learned someone did a third-party report on my behalf. I didn't want to participate at all, and I thought I would be left out of it. I didn't want to be the reason another person got discharged from the military."

"We need your testimony to clear this up," he came at me again. "The only way you and all the other victims will get justice is by your testimony. You were the only victim who had no personal connection to Lt. Lewis except for when he assaulted you."

He informed me that Lt. Lewis had lied about our "encounter" and said the reason he wanted to meet with me was because he heard that I was doing "black market deals" with government property, and he wanted to score a knife. Since he made an allegation of me stealing government property, I needed to participate.

The prosecutor then offered a win-win.

"Would you be willing to go this far with me?" he asked. "Testify up to the point where you feel uncomfortable. Then stop. You don't have to go into him penetrating you with his fingers. I want to protect as much of your restricted report as I can."

"Can I just say that I don't remember?" I asked.

"No," he said quickly. "You can't say that. You're going to have to talk all the way up to that point. If they ask, you need to tell the truth. But for my part, I'm going to do my best to ask questions that will keep us from going down that road."

I agreed to tell as much as I could tell, and they scheduled my deposition over the phone while the prosecutor and the others were in court in Southern California.

At the deposition, I told them what happened, including Lt. Lewis licking his lips, what he said to me about my body, pushing me against the desk, then telling me that I was *no fun* before he left.

The defense pushed for more details.

"How long had you known him before this happened?" he asked.

"I didn't. I'd never seen him before that day," I answered.

"Did he grope you when he pushed you against the desk?"

"His body came in contact with mine. That's all I remember," I said.

Repeating—reliving—this trauma made me feel broken, and I had come a long way in my healing. But I refused to go lower my head.

S.Sgt. Campbell later told me about his own testimony at the deposition. He'd been called since he spoke to me only moments after the assault took place.

"It got ugly, Kia," he said. "I mean, the defense really went after you when you weren't testifying. They slut-shamed you, asking me what you were wearing and all sorts of shit. They asked me how often you flirted with men, how openly you talked about your upcoming divorce from SJ. I mean, they threw you under the bus, backed up, and tried to hit you again."

Big surprise.

"If I didn't know you," he said. "Like, if I were sitting there having to pass judgment on you without knowing you like I do, I would have bought their shit. Now I see why victims of assault keep their mouths shut. This shit is traumatizing, and I was only a witness. I can't imagine what you all go through. I've always repeated the Air Force line of 'See something, say something.' But after this, I will always respect a person's wishes if they don't want to say a word."

S.Sgt. Campbell is one of the best people I've had the privilege to work with in the Air Force. He protects female airmen from predators, and he has become a strong advocate to help them.

At the end of the trial, despite the games the defense played trying to exonerate their defendant, Lt. Lewis was discharged from the Marines for his assault on me and the other victims. Even if I had told everything he'd done to me, I doubt he would have gotten a more severe punishment. As it was, he lost all pension and benefits, too. It took over two years to bring this chapter to an end, but I felt enormous relief. Lewis might continue to harass women in his civilian life, but he can no longer use his rank to prey on military women.

I don't have a vindictive bone in my body. I took no personal satisfaction from Lewis's discharge. But I made a stand, refusing to be used by anyone again. My old strength and courage were returning—little by little.

But I still had one unresolved issue: what should I do about SJ's sexual assault against me?

Chapter 31

If Not for Me, for My Daughter

"It is never wrong to do what is right."

~ Gift Gugu Mona

B Y THIS TIME, I had started seeing a therapist to help me deal with my trauma. I found that the process of talking through the abuse I'd endured, on all the various fronts in my life, helped liberate me. To have someone listen to me without judging! The more I confided in my therapist, the more I remembered *that strong woman who wanted to change the world for the better. That was me!*

Around the same time, the Family Advocacy Treatment manager informed me that I had one year from the most recent incident to file a complaint due to the statutes of limitations for intimate partners; once that year was up, there was nothing I could do. I needed to make the decision if I should file charges against SJ, and I needed to make it quickly.

I thought through my entire relationship with SJ, from start to end.

When I first met SJ, I'd just come out of a terrible relationship with a party-boy who lived to drink. Maybe that set my pendulum swinging too far in the opposite direction. Whereas Chris was just a boy, SJ was older, and he had earned respect from command. SJ felt safe, if maybe a little boring. And he played the role of the perfect gentleman—wooing me by opening doors, giving me flowers, and taking me to nice dinners.

Then he changed. But then I learned, not really. Along the way, I reached out to his "evil ex-wife," the one who supposedly cheated on him. SJ told me once that she tried to leave the house after he told her not to, and he told me how hard he punched her. He then adjusted to say that he had done so as self-defense. I knew why he shared that story with me: he was letting me know he would do the same thing to any woman who crossed him. His ex-wife told me how sorry she was for my problems with SJ, but she could offer no advice.

I thought back to when I'd been such a fun person, even a little feisty. Not only did I stand up for myself, but I also fought hard for those around me. As SJ and I had problems, he used my forgiving nature against me, pushing me further with each offense. When he got mad, he punched walls. But I feared it wouldn't be long until he hit me instead.

Throughout our marriage, I kept a journal. Initially, I wrote things down so I could reflect on how far I'd come. Eventually, I burned them all. But I found a few entries I'd written in a work journal that were spared the flame.

"It sucks to have to write this in my work notebook, but if he ever finds this, I'm sure I'll be a dead woman. If not dead, maybe I'll be beaten. He hits things around me but never me. One day he'll miss the wall and hit my face."

Then I found what I wrote when I first considered leaving him:

"He says I can never leave him, or I'll be condemned! God will never forgive me. He said that divorce is the biggest sin. So I guess I have no choice but to stay and learn to cope."

I had been so innocent, so naïve when I married him at age nineteen, yet I thought I had found true love. Nineteen. I wish my mom could have read my mind then and said something to challenge my delusional thinking. Maybe she could have told me, "You think you know everything cuz you're nineteen? You smellin' yourself?"

But once I entered marriage, I accepted SJ as the leader of our home. I listened to him and trusted him for the longest time. When he'd tell me "the Bible says," I'd just accept it as fact. When he yelled, I tried to diffuse his anger; when he cried, I offered comfort.

Over time, I would rage against his rage. Then I'd find myself screaming and playing the part of a lunatic, and it would be SJ who tried to calm me down. He so easily slipped into the victim role, like I was the monster. And the worst part of it was, I let him.

The more he acted like he was the boss of me, I considered him my boss—not a friend, lover, or husband. He had "outranked" me our entire marriage, using both tears and rage to gain control of me.

As I thought about how I'd allowed myself to act so out of character in our marriage, I felt self-loathing grow inside me. No one has the power to cause another person to lose his temper, but I would at times. I could work myself up into a righteous tizzy. Then I'd hate myself even more.

Eventually, I stopped fighting. I just let him be right, and I kept my mouth shut. I told myself, *If you're not going to do something about your situation, then stop complaining.*

I made my bed, so to speak. I could choose to reengage in this endless war at home, or I could choose peace.

I recalled the many people who'd told me throughout the years, "You've changed so much!" or "You're not the same since you got married." Instead of trying to understand what they meant, I got defensive, believing they were attacking SJ. Wanting peace, I would not say anything bad about SJ, and I wouldn't let anyone else, either.

Then I remembered the many times SJ cried after he hurt me, saying things like, "I'll do better" and "No one will ever love me like I love you." I believed those words even when his actions showed the opposite. I believed, because I needed to believe; to believe otherwise would reinforce that I would live in hell for the rest of my life.

I thought about the irony of how he loved me. At a mall once, a guy pulled SJ's arm off mine and put his own arm around my waist, saying, "I know a fine lady like you isn't with this dude. You need a real man, honey." SJ did nothing. I had to stand up for myself. Another time we volunteered to work concessions on behalf of the Air Force for a Tampa Bay Rays baseball game. The manager of concessions felt me up right in front of SJ, who watched it happen. Instead of stepping in, SJ turned his head. I ripped the manager a new one. Later, SJ yelled at me, "What did you want me to do? Get into a fight with the guy?" *My bad. I picked a bad time to be groped.*

As my heart grew heavier, I chided myself for dwelling on the darkest moments, forcing my thoughts to all the good times. *Like Tahiti! That was the best trip ever!* I conquered my fear on that trip and swam with stingrays and black-tipped sharks. It was amazing! But oh yeah, SJ, apparently, had his own fears about those sharks. When a shark got too close to him, he pushed me between himself and the shark—while he ran to the beach. When we both get back into the villa, he blamed me. "You're the one that planned this fucking trip!"

I often played referee between SJ and my mom and sisters. When my sisters yelled at me for choosing my husband over Mom, I couldn't explain myself. I'd ceased to be myself. Instead, I'd hear myself making excuse after excuse for him.

Any outsider hearing parts of my marriage saga would say, "I would never let anyone treat me like that. Damn, have some self-respect." But I didn't go from confident to doormat in one giant step; rather, my identity disappeared slowly, and over time—until I no longer recognized myself at all. The only time I'd realize how much I'd changed is when I put thousands of miles of distance between SJ and myself, and I could feel myself taking full breaths again.

Abraham Lincoln's "Emancipation Proclamation" announced that slaves were set free, but it took years for them to experience anything resembling freedom. My chains were less obvious than those of a slave, but they were no less restricting. And like Lincoln's proclamation, words don't change realities. Actions do. My divorce decree hadn't set me free. It was only a start. I needed to set actions into motion to secure my freedom.

In contrast to SJ, James treated me like a lady and encouraged me to live my life so that I could grow into the woman God made me. I want to someday be the best wife, mother, daughter, and sister—and those desires inside me had nothing to do with SJ.

Still, I had to decide what to do: should I press charges, or should I let it go? As I processed my trauma, I realized the only way to stop feeling like a victim was to act like a victor.

"I'd like to talk to you about my ex-husband," I told my treatment manager after I had replayed all the mental tapes from our marriage. "I'm going to file for sexual assault."

"Okay," she responded. "Let me get the proper forms out so we can begin."

I knew from the beginning that I faced an uphill battle reporting assault against my ex-husband. But I knew I had to try.

Early on, I shared the details with an officer.

"So, let me see if I got this straight," he asked me. "SJ got you drunk and sexually assaulted you last fall. But prior to that, he raped you?"

"Yes, sir," I replied. "That's correct."

"When exactly?" he asked.

"I have it written down and can give you the exact dates," I told him. "It happened more than once. At some point, I stopped keeping track because it became so frequent."

"Did you ever report this to anyone?" he followed up.

"No, sir," I told him. "SJ told me that he had the right to my body. Eventually, I stopped fighting."

"So you consented?" he asked.

"No, sir," I tried again to explain. "I mean, not at first. You have to understand. The first time he raped me, I was on leave for a back injury. I could hardly move. I couldn't even dress myself. So he carried me from the shower, and into the closet to help me get dressed. Then…" and I told him what happened.

He took several notes while I talked, but he didn't interrupt. Finally, he pushed back from his desk and took a deep breath.

"Like I said," I told him, "SJ told me what he did wasn't rape, because we were married. But I know that can't be right. I did not consent. He beat me down, and I was helpless."

"Let me be honest with you," he told me. "And let me preface this by saying these are not my personal opinions on the subject. I'm going to speak from the legal side of things."

"Okay," I nodded.

"I'm sorry for what you've been through and the pain you are feeling," he said. "But the textbook definition of coercion without threat of life or weapon is not rape. So on that one point, SJ is correct," he said.

Before I could interrupt, he held up his hand.

"And the second thing, something I'm sure his lawyer will bring up, is that he raped you, or whatever, but you stayed with him. Since he abused and raped you in the past, and you continued to stay without filing charges or notifying anyone," he paused while he found the best way to finish, "well, his lawyer will argue that SJ didn't know that him pouring you a drink and taking advantage of you was wrong later. Does that make sense?"

I had been in the military long enough to know that convoluted, counterintuitive mumbo jumbo means you're fucked.

"Let me make sure I understand this," I said, trying to follow his logic. "If I murdered twenty people, and I never got caught, how would I know that murdering number twenty-one would be a problem?"

"What I'm trying to say is that these types of cases are too hard to win," he answered. "If you say you were raped, yet you stayed each time, how was he supposed to know that this was the final straw, that 'raping' you was wrong when it was a long-established pattern? He rapes you, you two argue, you stay, it gets calm, he rapes you again, the cycle of abuse repeats. I'm not saying I agree with it, but I am telling you what his lawyer will claim on the stand. But I will work with you to move this forward."

Things moved very slowly until August 21, 2020, when this case came before the wing commander serving as the convening authority back at my base in Florida. While the prosecutor did not believe he had enough evidence to take this up with the court system, I had this one chance to speak at a telephone hearing before the wing commander to try to get some justice. Here's some of what I said:

> "My case may not look like the textbook definition of rape—sexual intercourse carried out forcibly or under threat of injury against a person's will—but it does not make what happened to me any less horrific and traumatizing. Because I do not fit the textbook definition

of rape does not make the nightmares go away. Because I do not fit the textbook definition of rape does not absolve what was taken from me. Because the law does not classify using religion, intimidation, and restraint to force someone to have sex as rape, I am left with picking up the pieces to try to come to terms with my rapist.

"Though the justice system fails to call him what he is, he will forever be engrained in my memories as the man who raped me repeatedly for years. My rapist isolated me from my friends and family. I was physically locked in a house for weeks only to be let out when he returned home. I had limited access to funds, even though I had my own income. My social media accounts were consistently under scrutiny, and with any new interaction I had, I was demanded to provide justification as to how I met them. I was not allowed to have my own car and had to rely solely on my rapist for rides to and from work. I was not allowed to have lunch at work with anyone else but him. I was not allowed to shave my legs or wear any revealing clothing (such as shorts, dresses, or tank tops) without his permission.

"When I had a crippling back injury and relied on my rapist to assist in daily activities such as getting dressed, walking, showering, and using the bathroom, my rapist took advantage of me. I will never forget the day we got into a fight because I did not want to have sex. He threw me down in my closest, sat on top of me, applying overwhelming amounts of pressure onto my already injured back, further numbing my legs. He sat on top of me and slapped his face until it was deep red, asking if this is what I wanted to do to him. Tears flooding my eyes with fear, pain worsening in my back, so that I found it so hard to breathe without pain. He began to shout Bible verses at me: "Wives shall not withhold themselves from their husbands and must submit themselves unto their husband as Christ submitted himself to the Lord." Over and over and over again, he repeated this verse and condemned me to hell, shall I break a promise that I made to the Lord. As he applied more pressure on me, the only way to make the pain in my body go away was to submit.

"That was the moment I was first raped. But the law does not see it this way, because no weapon was used against me to gain my compliance. Once my rapist was finished with me, he put me back in the bed and left to the kitchen, leaving me broken, sobbing, in pain and great shame. He later told me not to mention this to anyone, because "they will not understand," and "it will ruin my career." Then he asked, "What kind of wife would send her husband to jail?"

"I found it impossible to stay but more impossible to leave. He would not let me out of this marriage safely. I had no one left, and my rapist knew that. None of his abuse ever met textbook definition. He walked that fine, thin, grey line. I've been so numb, so conditioned that I stopped fighting it, because the more I fought the worse it was.

"I had found a new strength, and it was at that moment I knew I had to leave if not for me, for my daughter. I tried to file for divorce, and it was always met with him saying "your life will be condemned" and "you will go to hell." He took all the money out of our bank accounts, leaving me no choice but to stay. Whenever I tried to leave, he reminded me that I had no money and would never make it without him. I had no credit established, due to him refusing to allow me to get a credit card. I had no car, no funds, no support system in place to run to if I left. I felt so alone.

"It wasn't until I deployed that I made a separate bank account and sent all my money into it. This was my chance to set myself up to escape. Once I returned home, I started my course for a new life. Finding new places to live, doing a budget, and figuring my life out. When he returned home, he quickly returned to the same person I knew. He gave me wine as he sat and watched me. Slowly sipping on his one beer as he kept my glass full. Once I was too intoxicated to remember much of the night, he took advantage of me. The law says that if someone is too intoxicated, that means there is no consent; but in my case it does not apply. Since I can only remember the beginning of my assault and not the rest of it. That does not make him innocent.

"During our marriage, I went to the police four times: Saint Claire County, Virginia; Hillsborough County, Florida; Fairfield Police Department, California; and Vallejo Police Department, California. But each time I was told by law enforcement, "since you are military, you have to take it up at your base." But the military justice system has failed me repeatedly. Again, I was on my own to swallow my pain and keep my suffering to myself.

"Since I do not have enough evidence to convict my rapist, I must find closure in this situation on my own. Justice is not found in my many counseling appointments. Justice is not found in my nightmares, in a new city, in a new state, in a new home, or even in a happy marriage. I cannot find justice by simply moving on. While my rapist walks away a free man with a big sigh of relief that he beat the system, I, the victim, am left crawling through each day, struggling to find peace in what remains.

"The Central Registry Board (CRB) determined that my rapist met criteria for sexual abuse, but did not meet criteria for emotional/mental abuse. If a victim is raped how does that not cause any emotional/mental problems. If the CRB had enough evidence to find that he met criteria for sexual abuse, why does the law not support this finding?

"At what point do we start to take accountability? If not this time, then when? If not me, then which victim? When will be the "this is it" moment before justice finds him?"

The convening authority sided with me! I won! Happy ending!

No. We don't always see justice in this world. And not every story has a happy ending. The convening authority gave SJ a slap on the wrist, which amounted to telling SJ, "Don't do it again." I didn't win, and SJ was never held accountable for what he did to me. But God works in mysterious ways, not always the ways we want.

I had wanted out of Florida and off that base. Specifically, I wanted the commander to allow me to retrain and transfer. He had refused. Then I was sexually assaulted. As a result of the assault, I was given an expedited transfer. Is that the way I wanted to leave? Hell no. But God works in mysterious ways. I wanted a divorce, and I wanted to set an example for my daughter, Reign, so she wouldn't allow others to abuse her. I lost the case. But I stood up for myself, said my piece, went on record, and am stronger because of it.

Yes, indeed, God does work in mysterious ways.

My healing continues today. I'm not done yet. Part of that process involves therapy. Another part includes helping other service members who feel invisible and voiceless. And yet another part is writing this book, where my closest friends and even some family members will first learn about what I've been through during the last decade.

But my healing won't be complete until I can honestly encourage my own daughter to join the military, and that will happen if the military does the right thing for service members, makes some changes to catch up with the civilian world, offers zero protection to predators, and protects male and female members from physical and sexual abuse—for as long as they serve our nation.

In case you're wondering, I have a few ideas on how to do that...

Chapter 32

To the Doubters

"The best way to show that a stick is crooked is not to argue about it or to spend time denouncing it, but to lay a straight stick alongside it."

~ D.L. Moody

I F YOU HAVE read this far, you might find yourself with some questions, challenges, or doubts about my story, and that's okay. In fact, I'd be surprised if you didn't have one of these reactions:

1. **Don't blame the military**. "Your story is sad, but bad stuff happens outside of the military, too. You can't blame the military for the things that happened to you."
2. **Your experiences are the exception, not the rule**. "Your story is an outlier. I'm sure other service members had great experiences. If anything, you had some bad luck."
3. **You're lying**. "If your story is true, I'll bet you cherry-picked all the bad parts and exaggerated them for the sake of the story."
4. **You wrote this to trash your ex-husband**. "About half of all marriages fail like yours did. Why did you go into so much detail? Is this to get back at your ex?"
5. **What can we do to fix this**? "My gosh, that's terrible. What do you suggest the military do to change that culture?"

Instead of dismissing critics or cynics, I'd rather address them by responding to each of these possible reactions.

Don't blame the military.

235

I'm not the only one saying the military has a sexual abuse problem. Even the Department of Defense recognizes it, as reported by the RAND Corporation (Morral et al, 2015). I don't want to lose you with too many numbers, but I think these statistics from the 2015 Department of Defense Sexual Assault Prevention and Response report (Protect Our Defenders, 2020) show the problem is real:

- 1 in 4 female active-duty service personnel has faced severe and persistent sexual harassment.
- Most victims of sexual assault were assaulted more than once.
- Eighty-five percent of sexual assault victims do NOT report the crime.
- 1 in 4 victims who did not report feared retaliation from their chain of command or coworkers.
- Sixty-two percent of women who reported sexual assault faced retaliation, most often from *superiors and commanders.*
- A third of victims of sexual assault were discharged after reporting, usually within six months of filing a report.
- Forty-four percent of victims were encouraged to drop the issue and forty-one percent said the person to whom they reported took no action.
- Nearly half of survivors were dissatisfied with their treatment by their chain of command.

Do bad things happen outside of the military? Absolutely. Sexual harassment, sexual assault, and rape take place at especially high rates when one of three ingredients exists: power, youth, and alcohol. Sometimes they exist together. But one is always present: power. Wherever you see sexual crimes, you will find someone holding power over a victim. This plays out in corporations, scouts, churches, cults, and households. Wherever a power gap exists, abuse is more likely to happen.

For example, Harvey Weinstein was an elite power in Hollywood. He used that power to sexually assault and abuse numerous women. SJ used his role as the "leader of this household," physical strength, and emotional manipulation to keep power over me.

What does the military use to keep order within the ranks? Power. It's not surprising that our military system—a system run on power, command, and control—would have high rates of sexual abuse.

Youth also plays a role in sexual abuse. When an older predator targets a younger victim, it's still about power. Youth becomes a factor, because it's easier to manipulate and control someone who is younger, more innocent, and lacking in experience.

But sexual assault also takes place in youth cultures. According to the Rape, Abuse & Incest National Network, known as RAINN, 54 percent of victims of sexual assault are within the ages of eighteen and thirty-four. Sexual assault takes place outside the military, but wherever it occurs, the victims are most likely to be young; the same age as teens who go off to college. Or join the military. And their perpetrators? They are often the same age. When you put hormones, opportunity, and ignorance in the same room, even the most innocent people can get hurt.

Finally, alcohol is often an ingredient in sexual assault. Do you remember I mentioned my first military boyfriend, Chris, the one thrown out for, among other things, underage drinking? When I attended those parties during my early days, more than half of the attendees were twenty and under. I know I wasn't the only one who spent my first overnight away from home when I joined the military. Many of the new airmen around me were away from home for the first time, trying to grow up too fast, and enjoying their freedom for the first time. It's no wonder that so many assaults take place on college campuses and military institutions. You have inexperienced youths experimenting with alcohol and even drugs. It's a recipe for trouble.

So can I blame the military? Yes, I hold the military accountable for creating a culture where bad actors and bad leaders often thrive unchallenged and undisciplined for their despicable actions as well as inactions.

Your experiences are the exception, not the rule.

I'd like to respond to that in a couple of ways. First, review the statistics from the Department of Defense. If one in four service women became victims of sexual assault, I am not the exception. If 62 percent of female victims experienced retaliation because they reported their abuse, I am not the exception. If 44 percent of female victims were encouraged by their superiors to drop the charges against their perpetrators, I am not the exception. If nearly half of the female victims were dissatisfied with how their superiors handled their complaints, I am not the exception.

But I wish to God I were. If I were the only female in the military who experienced such traumas, I could at least take some solace that others thrived. I could even tell my daughter when she's older that she should follow in my footsteps to serve our country, because "all the bad people took out their aggression on me, so you'll be safe." But I'm not alone.

Sharing my story—exposing myself to further harassment, criticism, and scorn—wasn't about me "making a name for myself." I would give anything if this were not even my story to tell. I never wanted to bare my soul and share my shame with others. The tipping-point for me had nothing to do with events going on in my own life, when I sat down to write; rather, it had to do with watching the news unfold of the discovery of the lifeless body of Army SPC Vanessa Guillen.

Before Army SPC Vanessa Guillen was murdered, she told her family that she was the victim of sexual harassment by an Army sergeant.

Sound familiar?

Other female enlisted soldiers filed complaints against the same man, but those complaints were dismissed. Later, SPC Guillen was murdered by a fellow Army soldier inside the armory where he worked.

Sound familiar?

Watching the news unfold, I knew that I would do anything—even if it meant losing my career—to prevent another member of the military from experiencing what happened to her, and what nearly happened to me.

When I told some fellow airmen that I was writing my story, several of them wrote me to share some of their own stories.

A former colleague named George said this:

> "My whole military career is a regret to me. Those were the darkest days of my life. Why do you think I partied so much? I don't even like to drink. I used it as an escape. I partied because I was hurting. I fell into a deep depression [but] never went to counseling. I thought about hurting myself. You were a prime example of why I didn't want to go into counseling. You opened up, and they isolated you. I knew how they would treat me (George and Lindley, 2020)."

The leading cause of PTSD in the military is not from combat; it's from military sexual assault. According to the inspector general's report in 2019, less than 40 percent of service members meeting the criteria for PTSD experienced combat, whereas 45 percent of women service members who were raped experienced PTSD. It was even worse for me. Sixty-five percent of men who were raped in the military experienced PTSD.

So were my experiences the exception, not the rule? No. The problems of abuse within the military are pervasive. Some end up like me; some end up even worse. And even those who complete their military careers without ever saying a negative word often leave with addictions, depression, and despair.

You're lying.

Maybe. I mean, I'm not, but I realize you can't see what I used to write this book: documentation.

Early in my military career, an advocate from the inspector general's office told me that I needed to write down everything. So that's just what I did.

Here's how a complaint to the inspector general works. You have sixty days from the alleged incident to file a complaint with their office. If they didn't find enough information from that ONE incident, the case gets returned as "unsubstantiated" after they conduct a "thorough" investigation.

In my case, when I first went to the inspector general, they told me I didn't have enough information. That's when I heard that advice: document everything. So I did. Each time I experienced mistreatment, harassment, bullying, or abuse, I wrote it down. This information eventually backfired on me. My commander stated that I was "evil" for "blackmailing" people.

I never blackmailed anyone. Blackmail is an illegal action and is treated as a criminal offense, and it includes demanding payment or another benefit from someone in return for not revealing compromising or damaging information about them. I neither needed nor wanted anything from those who mistreated me, except to be treated fairly and for the abuse to stop. I documented events so no one could ever say, "You just now made that up." Then when I wrote things down, my commander called it "blackmail." Again, lose-lose situation.

When I told the judge at the conclusion of SJ's sentencing about the four times I'd spoken with civilian police officers, I didn't pull that out of the air or from my head. I pulled it out of my notes. In fact, I still have the cards those four police officers gave me in case I needed anything else. I have copies of every letter from home I received, every letter of reprimand, letter of counseling, and every commendation. On top of that, I kept a journal on my computer where I wrote short summaries of significant events as they were happening. And I have copies of the counselor's notes from when SJ and I went to sessions together.

Did I exaggerate? No. The only fictionalized accounts in this book involve two arenas: changing names and downplaying actual events. As far as names, I didn't write the book to commend or condemn anyone in my story. The only real names I used in this book are mine and James's. As far as downplaying events, I didn't do this to avoid embarrassment. I'm beyond that. But I'm not beyond being retraumatized. In a couple of stories, I gave you the PG-13 version when the actual events were R or NC-17 due to graphic violence. For the sake of my mental health, I did not wish to relive those times. I've worked too hard to get to a good place. I won't risk losing that progress.

I didn't share everything. If I had, the book would be longer than any one of Winston Churchill's numerous autobiographies. I had to leave some things out. Let me give you an example of the kind of things I didn't include, and you decide if I should have added it.

I got reprimanded for wearing bright red lipstick at work. But I wasn't wearing lipstick. I was "wearing" the remnants of the Slurpee™ I sucked on before going to work. My lips weren't the only red thing on me. My tongue and teeth were also colored from that wild berry concoction. But my supervisor didn't believe me.

"If that's the case, I need you to show me the receipt for the Slurpee™."

Okay, now let me ask a question to fellow snackers. Let's say you go into a 7-11 to buy a soft drink and a beef jerky. How long do you keep that receipt? Or maybe a better question is, if the clerk asks you, "Do you need a receipt?" do you say yes? Think about that. Do you take that receipt and hold onto it, fearing that at some point in the future, someone is going ask you to prove that you did, indeed, buy a beef jerky and a soft drink on that particular day?

So no, I didn't save the receipt. When I told my supervisor that I hadn't kept the receipt, she added to my reprimand for wearing lipstick that I lied about it as well.

Irritating? Absolutely. In fact, just remembering that makes me both angry and hungry for a beef jerky and a Slurpee™. But no, I tried to leave out stupid incidents like that, because those things happened regularly and didn't seem worth sharing.

Did my events in my story really happen? Yes. Any missing details were intentional to spare me—or the reader—from overkill.

You wrote this to trash your ex-husband.

My story is called *War at Home*. I had traumas wherever I called home at different parts of my life. My decision to share details about my marriage to my ex did not come quickly or easily.

Knowing that written words stick around, my daughter will read this someday. I wrestled with her knowing all these details. But in the end, I chose to record the events as they happened. I have worked diligently to avoid saying negative things to my daughter about her father. She knows that. She's seen him scream at me many times, but I have tried to foster the two of them having a close, loving relationship.

Then, I thought about all the other little girls who may grow up and make bad relationship decisions like I did. Since so many marriages end in divorce, I wrote to share my red flags in hopes that others can use it as a guide of what to avoid in their own relationships.

Finally, I considered all the men and women caught up in domestic violence who feel trapped, the ones who feel like no one cares about them or understands what they are going through. *You're not alone.* I had to learn so many lessons the hard way. But even as a married woman with a child who had gotten sucked in by a manipulative man, I found a way to get out. You can, too.

So no, I debated in my heart a long time before I wrote about my marriage. Even as I tried to edit what I wrote, I had to recognize that part of me still tries to protect him and his image. But trash him? God, no. I shared what happened not to hurt him but in hopes of helping others.

To reiterate, I did not share my story for attention, money, or sympathy; I became willing to be vulnerable in my telling of my personal experiences to help others. Which leaves me with the last, most important part of this book—

What now?

Chapter 33

What Now?

What can we do to fix this?

I'M SO GLAD you asked, because that is the reason I laid myself bare, sharing details of my life that I never wanted exposed. I want to be the voice for change, not the tears of injustice. I don't wish to bad-mouth the Air Force or any military branch. But I think we can do better. I think the women and men in the military deserve better.

When I leave military service, I aim to spend my time elevating these issues and pushing for change. Ultimately, nothing would make me happier than to have my own daughter join the military someday; but I couldn't imagine that happening under some of the current practices.

Here are six simple ideas that I believe will make a world of difference to those men and women serving our country:

1. **Allow service members to use outside counseling services without notification to the command.**

Today, service members use military counselors who notify command only if there is a potential "readiness" hindrance, such as suicidal or homicidal thoughts. Some believe that military counselors have deeper experience in dealing with situations that arise out of military life and service. For example, they see more cases of PTSD than their civilian counterparts. And it's believed that command has the right to know who is seeking help.

But it's not enough, and it doesn't work.

First, service members don't trust military counselors. Why? Military counselors communicate with the command. They are viewed as an extension of the military. There's a reason why priests don't moonlight as policemen. When a service member is willing to ask for help, they don't need to worry that their counselor is simply an extension of their military commander.

Second, those who need help won't ask for it because it comes at a high cost: loss of reputation and trust from their leaders. Even if the only information the counselors can share is if a service member poses a threat to themselves or others, the fact that they notify command is enough to keep them from help. Remember, according to the Department of Defense, victims reporting abuse receive retaliation much of the time. What do you think happens to the career of a service member who needs help when the very act of seeing a counselor can cause their supervisor to believe they are "unstable," "weak," or "unreliable"? Do you raise your hand for help when you know it will negatively impact your career path? Do you really want to see a counselor knowing that certain career doors close to you if you *ever* sought mental health counseling? My friend, George—one of many I spoke to while writing this book—is an example. I am another example. And we are not alone. Service members are taught to be tough. Too many suffer alone.

Third, problems that aren't dealt with while a service member is on active-duty will create a logjam of cases for the VA later—a system that is even less equipped to deal with volume. According to the Department of Defense's 2015 report cited in the previous chapter, over 1.3 million outpatient visits to the VA took place because of Military Sexual Trauma (MST)-related care in one year! According to their own reports, 38 percent of female personnel and veterans have experienced MST.

The VA is based on funding. If funding is low, then veterans are the first to suffer. If military members felt safe enough to seek treatment while still active, they would be covered by Tri-Care (the health care program for uniformed service members, retirees, and their families around the world) to seek outside providers and or utilize the free mental health clinic, which could drastically reduce the pressure on the VA hospitals. This alone could reduce the number of people needing PTSD treatment from the VA.

Service members need counseling services for other reasons than sexual abuse. We often hear reports of active-duty military members dying in combat from improvised explosive devices. Others die in training accidents. But do you know what the second leading cause of death is for active military members? *Suicide.* Read that again. Each day, twenty veterans and active-duty service members take their own lives, a number much greater than the general population. According to the 2018 Department of Veteran Affairs report (VHA, 2018) on mental health and suicide prevention, suicide rates for veterans, in fact, are 1.5 times greater than for civilian adults after adjusting for age and gender. Additionally, nearly half of all service members who died by suicide had known mental health conditions. The military needs to make it easier for service members to receive independent, confidential help to stop this crisis.

Finally, untreated mental health issues cost lives. When a service member struggles with a mental health issue, they put themselves and others at risk. I can't tell you the number of fellow airmen who told me they thought of hurting themselves or others. I wrote about a few of them. For each person I helped get the support they needed, there are many others who won't risk their careers by admitting they need help. I mentioned suicide, but military personnel are also more likely to struggle with addictions. A RAND Corporation study on substance abuse published in 2018 found that active-duty personnel participated in binge drinking 5 percent more than their civilian counterparts (Meadows et al., 2015). Additionally, active duty and veterans experience higher rates of Substance Use Disorder, depression, PTSD, homelessness, and as mentioned, suicide.

The stress of military life both directly contributes to mental health problems and indirectly causes the problems to fester. As angry as some taxpayers get about military spending, most Americans recognize the need for a strong military. But our strength comes from more than hardware. Our true strength is the people holding the hardware. If they are not mentally strong, we are all at risk.

2. Third-party reporting by an independent entity outside of the military.

Today, all reporting takes place in-house. If a problem takes place on the base, the military handles it on the base. Many times, even when problems take place off base, civilian authorities won't touch it, deferring to the military to handle matters how they best see fit.

The problem is that this doesn't usually result in justice taking place. While stationed at Fort Bragg, US Army Captain Erin Scanlon was sexually assaulted by another soldier at a party. She filed charges. As a result, here's what she had to say happened to her:

"My behavior was in question, my character as a person, and as an Army officer, were in question, and it shouldn't be like that."

After her experience, now-retired Captain Scanlon advocates that reporting, investigation, and rulings should be handled outside of the military.

"…[I]t's impossible to get a fair investigation when you're investigating yourself, because it is a family." Do you recall me sharing that more than one leader in the Air Force told me essentially that I should, "Keep it inside of the house, the family. Don't air our dirty laundry to others." What happens should you try to take problems outside of the family is that they close ranks around you. They twist the situation around to you, making *you* the source of the problem. More than once, I was slut-shamed for being assaulted. My cries for justice were met with "boys will be boys" and even "girls will be girls." Many times, my leaders told me to "suck it up" and "if you can't handle a little teasing, you won't make it in the military (Lozano, 2020)."

Instead of me and other victims being told we "won't make it in the military" if we resist being objectified, harassed, touched, and violently penetrated, the current role the military plays in serving as police, investigators, judge, and jury should be removed from their hands and placed elsewhere—somewhere that will do the necessary work to protect those that protect our nation.

3. Independent investigation.

Currently, each military branch has its own internal investigative branch. In the Navy and Marines, it's the NCIS. The Air Force has the OSI. In the Army, it's the CID. The skills and depth of investigations relies on the individual investigators.

But the current system has multiple failures baked into it. As reported in the December 9, 2020 *Army Times*, the so-called military investigative experts are "largely inexperienced, under sourced, and understaffed (Rempfer, 2020)." That was the key takeaway from an outside group of evaluators comprised of five civilians about the CID agents responsible to investigate felonies and military law violations at Fort Hood, the base where Vanessa Guillen was murdered.

The report found that over 92 percent of the enlisted investigative agents were fresh from a sixteen-week training course and not fully accredited to conduct investigations without oversight.

What's true of the Army and the CID is likely to be found true of the NCIS and OSI in the other branches. You can draw your own conclusions about why internal investigations seem such a low priority in the military.

I recommend that the military uses an independent investigation group that does not directly report to the DOD. Each service member deserves a team of skilled, highly trained, certified investigators. To date, the military hasn't demonstrated that internal investigations are their priority.

4. Remove the responsibility of commanders to determine rulings and judgments on cases.

Imagine you work at Macy's in the men's shoe department, and while you are working an employee from men's suits gropes you. Who would make a ruling on what should happen? Well, if you're in the military, the store manager would decide the outcome.

That's right. In the military, the base commander is known as the convening authority—a person not trained in human resources or law—makes the decision. And just like what would happen in the hypothetical Macy's example, can you see getting justice if both employees work for him? Think about you in your own job. Were you required to pass judgment on a situation involving two of your employees or coworkers, wouldn't you have some bias, either intentionally or not? The same is true of a base commander who has several layers of leadership as his or her eyes and ears. The rumor mill takes over, making whatever you've heard most often, most recently, or from your most-trusted leader mask impartial, unfiltered truth. If one of the parties involved is actively involved and highlighted for their work or volunteerism, then that can also play judgement in their rulings.

In October of 2020, the Pentagon released the findings of a three-year exhaustive study of all military branches that involved penetrative sexual assault cases closed between Oct. 1, 2016, and Sept. 30, 2017 (Britzky, 2020). Here are two eye-opening findings of that study:

- Seventy percent of the cases resulted in commanders taking *no administrative, nonjudicial, or judicial action.*
- Less than 5 percent of the cases resulted in a conviction.

No one suggests that the conviction rate should reach 100 percent. That's not feasible for a number of reasons. But failing to convict 95 percent of the time? That's insane.

People build relationships in the military. That's a good thing, until it clouds judgment. SJ had so many members of his leadership team supportive of him, they tried to stop our wedding because they feared I would pull him down! An S.M.Sgt. in SJ's office lied for him during an official investigation to help SJ get out of trouble. If that's not favoritism, bias, and a complete breakdown in the justice system, then I don't know what is.

Real judges, though, are trained lawyers. They understand the law, and they recuse themselves from cases where they can even be perceived as having a conflict of interest. That doesn't happen in the military. Most of the time, the commander passing judgment is from the base of the accused and the victim.

Commanders hold a bias, if not consciously, unconsciously. They must choose between a plaintiff and the accused, and most of the time they will go into passing judgment based on the rumors they've heard and the reputation of the parties, instead of the evidence in the matter. Additionally, playing the judge is just one more of the "other duties as assigned."

In 2019, NPR reported the story of freshman Senator Martha McSally of Arizona, who revealed in a Senate Armed Services subcommittee hearing that she was a military victim of sexual assault while serving in the Air Force (NPR, 2019). Like too many others, McSally chose not to report the abuse, not trusting the system to handle the matter appropriately.

In response, Sen. Tammy Duckworth of Illinois, a fellow combat veteran, made these remarks (Duckworth, 2019):

> "…[T]he military has utterly failed at handling sexual assault through the Uniform Code of Military Justice process, and I will push for meaningful reforms. As a former commander of an assault helicopter company, I want to know what else can be done beyond successful prosecutions that bring perpetrators to justice to make the lives of survivors better and ensure they have what they need to heal and be able to resume the careers they dreamt about from the time they entered the military."

I don't believe that training commanders to do a "better" job is the answer. Even if they doubled their successful prosecution rate of sexual perpetrators to 10 percent, it's not enough. Besides, judges have attended law school and passed the bar. It would be a waste of time and resources to attempt to educate commanders to make slight improvements in what is currently a small portion of their jobs. Instead, the military needs trained professionals in place to ensure that each case is heard and decided fairly.

5. **Establish a human resources department for incidents that do not meet inspector general or equal opportunity criteria.**

Human Resources needs training and a mandate to get involved in conflict resolution. Yes, the Air Force and other branches of the military have an HR department. Like civilian HR departments, military counterparts help navigate promotions, performance evaluations, separations, retirements, benefits, entitlements, retention, awards, career pathing, decorations, retraining, and personnel development programs. What's missing? Conflict resolution.

In the civilian world, HR specialists and generalists spend a great deal of their time asking questions and listening to answers. In many cases, HR doesn't need to do *anything* except hear a person out. In the military, sharing with a superior—even when it's done under the umbrella of protected communication—has a way of coming back on you. Human resources professionals have the training to know what situations must be dealt with, as well as which ones fall into the category of letting someone vent or serving as a sounding board.

Many cases that get escalated into filing reports could be avoided if service members had a non-biased, professional person to go to for guidance. Additionally, certified HR professionals receive years of training in critical thinking, ethics, behavioral science, negotiating, coaching, counseling, and other skills that are often lacking in military leaders who came up through the ranks.

6. Give the reporting party the right to choose between using military or independent party investigators.

Allow reporting parties to shop around for those they wish to represent them. In the civilian world, both plaintiff and defendant get to choose the attorney they wish to work on their behalf. Why should it be any different in the military?

Not all attorneys and prosecutors have the same skillset and passion for their work. The free enterprise system pushes the best ones to the top. Having been the victim of crimes committed against me while serving in the military, I learned firsthand that military prosecutors turn away from cases they don't believe they can win. For example, the findings of the Criminal Investigation Command (CID) report from the Army around the murder of Vanessa Guillen showed the investigators were "highly inexperienced" and "overworked." These factors led to her case being grossly mishandled, and investigators may have been morally (if not legally) complicit in her death.

Military investigators only have the time to conduct surface-level investigations. Imagine if a victim could bring in their own experts as part of the investigation instead of needing to have photos, eyewitnesses, and active "smoking guns" before any action takes place.

As a plaintiff, I would be more satisfied that the military did everything that could be done for me if they allowed me the benefit of neutral, trusted experts to look out for my welfare, instead of the interest of the chain of command.

Call to Action

If you are a member of the military suffering from hazing, sexual harassment, or abuse of any kind, contact your members of Congress. If you're concerned about possible retaliation, reach out to your congressmen through social media anonymously. This channel gets action. Little happened in the Vanessa Guillen case until news exploded, thanks to outcries on social media.

For both military members and civilians alike, let your voice be heard. Contact your congressmen and educate yourself on the I Am Vanessa Guillen Act. Ask your members of Congress to pass this into law immediately. (For an important update on this legislation, please see the Note to Readers in the front of this book.)

Service members know they might be called to go into life-threatening situations when they join the military; however, no one expects or deserves to be *killed at home* and *by the hand of someone inside the military* because of their service.

A Special Word for Victims of Assault or Abuse

If you are in an abusive relationship—or wonder if you may be in one—you might feel completely alone. Abusers are skilled at keeping victims isolated, removing any contact a victim might have with friends or family. Without a neutral, supportive sounding board, victims start to wonder, like I did: *Am I the crazy one here?*

I want you to know that you are not alone. This is not only your war, but this is my war, too. It's the war of too many women and children—and even men—across the globe. I didn't want to admit my abuse, either. Many people in our society place a stigma on those that report abuse, either shaming the victim, minimizing the abuse, or denying abuse even occurred. None of that means abuse doesn't happen; it just makes it less likely for victims to talk about it and get help.

But help is available. Visit www.rainn.org to talk to or chat online with a trained professional today. If you're not a victim of abuse or assault, the RAINN organization can provide you with research and resources so you can help others who feel trapped in their situations.

Epilogue

CURRENTLY, I'M FINISHING up my military career. I chose not to re-enlist. After reading my story, I don't think I need to say more.

As I transition out of the military in the next couple of years, I'm eager to start my next career—one where I will be able to help women who've been in situations like my own. I plan to speak to military members across the country on domestic violence, sexual harassment, resilience, and how we can collectively make the military safe for all members willing to serve this country.

References

"Mental Health." Veteran Suicide Data and Reporting – Mental Health (Va.gov), US Department of Veteran Affairs, Nov. 2020, www.mentalhealth.va.gov/suicide_prevention/data.

Morral, Andrew R., Kristie L. Gore, Terry L. Schell, Barbara Bicksler, Coreen Farris, Bonnie Ghosh-Dastidar, Lisa H. Jaycox, Dean Kilpatrick, Steve Kistler, Amy Street, Terri Tanielian, and Kayla M. Williams, Sexual Assault and Sexual Harassment in the U.S. Military: Highlights from the 2014 RAND Military Workplace Study. Santa Monica, CA: RAND Corporation, 2015. https://www.rand.org/pubs/research_briefs/RB9841.html.

"Military Sexual Assault Fact Sheet: Protect Our Defenders." Protect Our Defenders |, 20 Aug. 2020, www.protectourdefenders.com/factsheet/.

"Scope of the Problem: Statistics." RAINN, www.rainn.org/statistics/scope-problem.

Department of Veterans Affairs, Veterans Health Administration, Office of Mental Health and Suicide Prevention. (2018). Veteran suicide data report, 2005–2016. Retrieved from https://www.mentalhealth.va.gov/docs/data-sheets/OMHSP_National_Suicide_Data_Report_ 2005-2016_508-compliant.pdf

Meadows, S.O., Engel, C.C, Collins, R.L, et al. (2015). Health Related Behaviors Survey: Substance Use Among U.S. Active-Duty Service Members. Santa Monica, CA: RAND Corporation, 2018. https://www.rand.org/pubs/research_briefs/RB9955z7.html(link is external).

Lozano, Michael. "'That's My Justice': Former Fort Bragg Soldier, Sexual Assault Survivor Advocates for Military Oversight." ABC11 Raleigh-Durham, WTVD-TV, 2 July 2020, abc11.com/iamvanessaguillen-pfc-vanessa-guillen-vanesssa-erin-scanlon/6290432/.

Rempfer, Kyle. "Fort Hood Report Highlights Army CID's Failings There, and Possibly Elsewhere." Army Times, Army Times, 10 Dec. 2020, www.armytimes.com/news/your-army/2020/12/09/fort-hood-report-highlights-army-cids-failings-there-and-possibly-elsewhere/.

Britzky, Haley. October 30, 2020. "New Pentagon Report Reaffirms 'Systemic' Failure In the Military's Handling of Sexual Assault." Task & Purpose, 2 Dec. 2020, taskandpurpose.com/news/pentagon-sexual-assault-evidence-report/.

"Sen. McSally Says She Too Is A Military Sexual Assault Survivor." NPR, NPR, 7 Mar. 2019, www.npr.org/2019/03/07/700998400/sen-mcsally-says-she-too-is-a-military-sexual-assault-survivor.

"Duckworth: Military Must Improve Handling of Military Sexual Assault." U.S. Senator Tammy Duckworth of Illinois, 6 Mar. 2019, www.duckworth.senate.gov/news/press-releases/duckworth-military-must-improve-handling-of-military-sexual-assault.

About the Author

FROM AN EARLY AGE, Jakia Lindley dreamed of being a judge, an attorney, or simply being part of something greater than herself. A native of Canton, Ohio, she joined in 2010 and has since built a distinguished career in logistics and operational support. Throughout her time in service, she has led and supported critical missions on behalf of global combatant commands and joint operations. She holds a Bachelor of Applied Science in Technical Management from American Military University, where she graduated cum laude. Her leadership and advocacy have earned her several prestigious honors including the Air Force-level NAACP Roy Wilkins Renown Service Member Award, the Blacks in Government Meritorious Service Award, and most recently the Spirit of Hope Award named in honor of Bob Hope. Jakia continues to use her voice to champion reform, resilience, and representation, serving beyond the uniform and leading with purpose.

Call To Action

War at Home is not just a book. It is the beginning of a movement.

Help raise awareness and initiate change by visiting

https://jmlindley.com to:

- Discover exclusive stories, facts and pictures that weren't included in the book;
- Share your own story of survival with Jakia;
- Get updates on Jakia's Activism activities and news about the book;
- Learn how to join the movement and be an agent of change.

www.ingramcontent.com/pod-product-compliance
Lightning Source LLC
Chambersburg PA
CBHW082144120626
46553CB00010B/2756